REFLECTIONS ON SPACETIME
Foundations, Philosophy, History

Edited by

U. MAJER

Institute of the History of Science, University of Göttingen, Germany

and

H.-J. SCHMIDT

Department of Physics, University of Osnabrück, Germany

Reprinted from
Erkenntnis
Volume 42/2, 1995

WKAP ARCHIEF

KLUWER ACADEMIC PUBLISHERS
DORDRECHT / BOSTON / LONDON

A C.I.P. catalogue record for this book is available from the Library of Congress.

ISBN 978-90-481-4612-3

Published by Kluwer Academic Publishers,
P.O. Box 17, 3300 AA Dordrecht, The Netherlands.

Kluwer Academic Publishers incorporates
the publishing programmes of
D. Reidel, Martinus Nijhoff, Dr W. Junk and MTP Press.

Sold and distributed in the U.S.A. and Canada
by Kluwer Academic Publishers,
101 Philip Drive, Norwell, MA 02061, U.S.A.

In all other countries, sold and distributed
by Kluwer Academic Publishers Group,
P.O. Box 322, 3300 AH Dordrecht, The Netherlands.

Printed on acid-free paper

TABLE OF CONTENTS

REFLECTIONS ON SPACETIME – FOUNDATIONS, PHILOSOPHY AND HISTORY

During the academic year 1992/93, an interdisciplinary research group constituted itself at the Zentrum für interdisziplinäre Forschung (ZiF) in Bielefeld, Germany, under the title 'Semantical Aspects of Spacetime Theories', in which philosophers and physicists worked on topics in the interpretation and history of relativity theory. The present issue consists of contributions resulting from material presented and discussed in the group during the course of that year.

The scope of the papers ranges from rather specialised issues arising from general relativity such as the problem of referential indeterminacy, to foundational questions regarding spacetime in the work of Carnap, Weyl and Hilbert.

It is well known that the General Theory of Relativity (GTR) admits spacetime models which are 'exotic' in the sense that observers could travel into their own past. This poses a number of problems for the physical interpretation of GTR which are also relevant in the philosophy of spacetime. It is not enough to exclude these exotic models simply by stating that we live in a non-exotic universe, because it might be possible to "operate time machines" by actively changing the topology of the future part of spacetime. In his contribution, Earman first reviews the attempts of physicists to prove "chronology protection theorems" (CPTs) which exclude the operation of time machines under reasonable assumptions. He formulates a new technical condition which is intended to capture in a more satisfactory way what could be meant by "time machine". But, Earman argues, the improved definition makes it much more difficult to prove corresponding CPTs. The situation here may be similar to that presented by Penrose's hypothesis of cosmic censorship.

The two contributions of Coleman and Korté (CK) present a new analysis of the epistemology of geometry and apply it to the conventionalist-realist debate. In particular, they aim to show how the physical geometry of space (respectively spacetime) can be determined empirically in a non-conventional manner. For this purpose, they propose the distinction between formal, theoretical and physical coordinates as well as a number of principles pertaining to the nature of the measurement of geometric fields. The metatheoretic notions of theory completeness and epistemic decidability are then defined and discussed. 'Theory completeness' is similar to concepts discussed by other philosophers, such as that of 'semantical consistency' (von Weizsäcker) or that of a 'complete theory' (Carrier). On the other hand, 'epistemic completeness' seems to be a new concept which imposes the additional requirement that there must be non-

Erkenntnis **42**: 121–123, 1995.
© 1995 *Kluwer Academic Publishers.*

1

circular criteria for the identification and the correct behaviour of the probes and processes used for fundamental measurements. In their second paper, CK apply these principles to the measurement of the Euclidean metric by rigid rods as well as the determination of the spacetime metric by light propagation and free fall. The latter is based on the approach of Ehlers, Pirani and Schild (EPS) which has its roots in the work of Kretschmann, Lorentz and Weyl.

In the philosophical debate in which they partake, CK concentrate on 'post-differential-topological criteria', that is, they take for granted the differentiable structure of the spacetime manifold. However, in the original EPS paper, this differentiable structure is itself constructed by means of so-called 'radar charts'. Yet EPS had to assume world-lines of freely falling particles as one-dimensional manifolds, primitive clocks, as it were. In his contribution, Schmidt now shows that this assumption is unnecessary in the sense that clocks can be constructed at a 'pre-topological' level following an idea of Castagnino.

A different, although related, approach to the axiomatisation of space-time theories was represented in our group by Schröter. For details of this approach, we refer the reader to the overview given in the volume *Semantical Aspects of Spacetime Theories*, edited by Majer and Schmidt, Mannheim (1994).

The paper of Mühlhölzer addresses the general question of the fixity of reference in the context of the change of theory from classical (Newtonian) to special relativistic (Einsteinian) physics. As a first approach, Mühlhölzer argues that there is constancy of reference in this case, making use of the notion of "conserved mass" introduced by Ehlers. If this analysis is adequate, not much remains of Kuhn's notion of incommensurability. However, there is a more radical stance that has been taken with respect to reference, namely Quine's and Davidson's thesis that reference is 'in-scrutable'. If this position is sound, Kuhn's thesis would lose much of its interest. But, argues Mühlhölzer, we still have the possibility of relative interpretation, which would save as much "referentiality" as we need.

Norton discusses a central principle of GTR, namely general covariance, and Einstein's interpretation of it. He reviews different attempts to give general covariance a physical, not just a formal, meaning and finds these attempts unsuccessful. In particular, Einstein's view of general covariance as a generalised relativity principle expressing the equivalence of arbitrary accelerated frames seems to be misleading. Norton partly explains the confusion about covariance and relativity as due to the lack of those mathematical methods available now which automatically solve some of the problems Einstein had to grapple with.

Like that of Norton, the papers of Friedman and Majer have a more historical colour, dealing with the reactions of philosopher-scientists like Carnap, Weyl and Hilbert to the rise of non-Euclidean geometry and its successful application to the physical world in Einstein's theory of relativity. Friedman shows that, in the case of Carnap and Weyl, the reaction did not consist of a turn to pure empiricism but rather an effort to modify and generalise Kant's original position. Both Carnap and Weyl connect their analyses of space(time) to Husserl's phenomenological idea of "Wesenserschauung", yet in rather different ways. This leads to quite different results. For Carnap, the full metrical structure of physical space can only be conventional, whereas for Weyl the mutual "orientation" of tangent spaces is empirically determined.

Husserl's influence was not restricted to Carnap and Weyl, but, as Majer points out in his paper, a different aspect of his phenomenology, namely the doctrine of the "Lebenswelt", played an important role in Hilbert's philosophy of geometry. According to Hilbert, geometrical knowledge has two sources, pure intuition in a Kantian sense and our experience with nature. These apparently conflicting views are reconciled by considering Euclidean geometry as approximately valid in the "Lebenswelt" which is then presupposed in making more precise measurements. These in turn form the basis of geometry conceived of as natural science.

It hardly needs to be said that the present articles do not contain final answers but rather intermediate results and proposals for further research. We hope that readers will find this selection informative and stimulating in their own deliberations about spacetime.

Finally, we would like to thank the ZiF for the financial and moral support of this work.

ULRICH MAJER AND HEINZ-JÜRGEN SCHMIDT
September, 1994

OUTLAWING TIME MACHINES: CHRONOLOGY PROTECTION THEOREMS

1. INTRODUCTION: TIME TRAVEL AND TIME MACHINES

At the outset I want to distinguish between two questions which are often run together: (1) Is time travel possible? and (2) Is it possible to operate a time machine? Each question can be treated in various contexts. The context I will explore here is provided by the classical form of the general theory of relativity (GTR), although towards the end I will make some remarks about the implications of quantum theory and quantum gravity.

In the setting of a general relativistic spacetime M, g_{ab} I will take the key necessary condition for time travel to be the existence of a closed future directed timelike curve (CTC).[1] To make time travel practicable, one might also want to place some restriction on the total acceleration $TA(\gamma)$ of such a CTC γ.[2] For if γ is the world line of a rocket ship which is supposed to accomplish the time travel journey, the ratio of the mass of the payload to the total mass of the rocket is governed by the inequality (Malament, 1985)

$$\frac{m_{pay}}{m_{pay} + m_{fuel}} \leq \exp(-TA(\gamma)).$$

If $TA(\gamma)$ becomes too large, the fuel requirements become wildly impracticable for pushing around γ a rocket with a payload capacity sufficient to accommodate a human passenger. However, I will not be much concerned here with matters of practicability as opposed to physical possibility. The main aim is to explore negative results, and the less that is required of time travel, the sharper the negative result.

With these understandings in place, one can now ask: Is time travel possible according to classical GTR? The straightforward answer seems to be "Yes!" For there are cosmological models M, g_{ab}, T^{ab} such that the spacetime M, g_{ab} contains CTCs, g_{ab} and T^{ab} jointly satisfy Einstein's field equations (EFE) and the stress-energy tensor T^{ab} satisfies standard energy conditions.[3] An example of an energy condition is the so-called *weak energy condition* (WEC) which requires that $T_{ab}V^a V^b \geq 0$ for any timelike vector V^a. The WEC assures that the energy density of matter and fields as measured by any observer is non-negative. The WEC may be violated for quantum fields (see Section 5), but it is assumed to hold for any physically possible classical matter/field. The assertion of the possibility of time travel is not upset by the grandfather paradox and related paradoxes of time travel, or so I have argued elsewhere (Earman, 1995).[4]

5

Erkenntnis **42**: 125–139, 1995.

However, this positive answer to (1) does *not* entail a positive answer to the question of whether it is physically possible to build and operate a time machine. Consider, for example, the Gödel cosmological model. In this model EFE (with non-zero cosmological constant[5]) are satisfied and the WEC is fulfilled. Furthermore, the Gödel universe permits time travel; indeed, it permits it with a vengeance: through every point in Gödel spacetime there is a CTC. Thus, there is no plausible sense in which the Gödel universe illustrates the operation of a machine that somehow brings about CTCs, for in this universe CTCs are everywhere and everywhen.

Here it may be helpful to distinguish a strong and a weak sense of 'time machine'. A time machine in the strong sense is responsible for manufacturing or bringing about CTCs. A time machine in the weak sense does not create CTCs but simply enables an observer to trace out a CTC. In the Gödel universe any CTC is a non-geodesic. Thus, in this setting a sufficiently powerful rocket engine (as shown by Malament 1985 it would have to be powerful indeed) that allows an observer to trace out a CTC would count as a weak but not a strong time machine. In the present paper I will be concerned mainly with time machines in the strong sense.

Intuitively, a strong time machine operates by manipulating matter and fields in some finite region of space so as to manufacture CTCs. For example, since the unique exterior field solution of EFE corresponding to a source consisting of an infinitely long rotating cylinder of matter contains CTCs,[6] it is conceivable that setting a finite cylinder into sufficiently rapid rotation would eventuate in CTCs. The idea of a chronology protection theorem (CPT) is to extract from the so-far vague description of strong time machines some precise necessary conditions and then to go on to prove that these conditions cannot be met without running afoul of some plausible physical constraints.

CPTs take the form of impossibility results. Such results have played an important role in the history of physics. Whether or not CPTs can play a role as important as, say, the impossibility of perpetual motion machines remains to be seen. But an initial reason for thinking that the issue of CPTs is central to an understanding of the foundations of GTR comes from the fact that CPTs are closely connected with Roger Penrose's cosmic censorship conjecture which has been deemed to be one of the most important unsolved problems in classical GTR (see Wald 1984, p. 303). The philosophy of space and time, which has long showed a fascination with time travel and time machines, also has a stake in CPTs. Philosophers have a contribution to make in formulating chronology protection conjectures; for, as will be seen below, some of the early attempts at CPTs suffered from the lack of a clear analysis of what is involved in a time machine.

In the next three sections I will explore attempts to prove CPTs within classical GTR. I will argue that none of the formal results that have appeared in the literature are successful in undermining the possibility of a strong time machine.

2. PRODUCING CTCS: TOPOLOGY CHANGE

If it is going to be said that a time machine creates CTCs where none existed before, then it must make sense to talk about a time at which the machine began to operate and before which no CTCs existed. One way to cash in this idea for a relativistic spacetime M, g_{ab} is to require that there exist a *time slice* Σ (i.e. a spacelike hypersurface without edge) such that there are no CTCs in the *causal past* $J^-(\Sigma)$ of Σ.[7] The first part of this requirement immediately rules out the Gödel model since Gödel spacetime does not contain a single time slice. If M is simply connected, then no future directed timelike curve can intersect a time slice Σ more than once. Thus, if M, g_{ab} admits a time slice Σ, then by passing to a covering spacetime if necessary, it can be assumed that Σ is *achronal* (i.e. is not intersected more than once by a timelike curve). An achronal time slice is referred to in the physics literature as a *partial Cauchy surface*.

Note how far we have come so quickly. Even before beginning to talk about the details of how a time machine might operate, we are led to the conclusion that the spacetime must admit a partial Cauchy surface, which means that the majority of science fiction stories about time travel are already ruled out; for turning on the time machine will not permit the would-be time traveler who starts on Σ to travel to the past in the sense of visiting $J^-(\Sigma)$. This is just a matter of relativistic geometry – no physical laws have been brought into play. Those who hanker after the time travel of science fiction may be sufficiently discouraged that they do not want to continue. To those, I bid adieu.[8]

My task now is to illustrate how a time machine can create CTCs to the future of Σ. One way the time machine might operate is to produce a change in the topology of space within a finite region, for then CTCs must result. In fact we have

LEMMA 1 (Geroch 1967; Hawking 1992). Let M, g_{ab} be a time oriented spacetime.

(a) Suppose that $K \subset M$ is a compact set whose boundary is the disjoint union of two compact spacelike three manifolds Σ and Σ'. If $\Sigma \not\equiv \Sigma'$, then K contains CTCs.

(b) Let Σ and Σ' be two (not necessarily compact) time slices. And let $T \subset M$ be a timelike cylinder which intersects Σ and Σ' in the compact regions Σ_T and Σ'_T respectively. If $\Sigma \not\equiv \Sigma'_T$ but the portion of the tube $T_{\Sigma,\Sigma'}$ capped by Σ and Σ' is compact, then $T_{\Sigma,\Sigma'}$ contains CTCs.

Case (a) covers a topology change in a spatially closed universe while (b) covers a topology change in a finite region of a spatially open universe. The idea behind the proof of Lemma 1 is quite simple. Since the spacetime is assumed to be time oriented, there is a C^0, non-vanishing, timelike vector field V on M. In case (b) V can be chosen to be tangent to the walls of the tube T. Construct the integral curves of V. Suppose that every such curve which departs Σ_T in the future direction reaches Σ'_T. Then there would be a continuous one-one map of Σ_T onto Σ'_T. Since Σ_T and Σ'_T are assumed to be non-homeomorphic, some such curve γ must fail to reach Σ'_T and must, therefore, wander endlessly in $T_{\Sigma,\Sigma'}$. Since $T_{\Sigma,\Sigma'}$ is compact, γ has a limit point p in this set. By modifying γ slightly on each close encounter with p, γ can be deformed into a CTC.

So provided that it is possible to manipulate matter and fields in a finite region of space so as to produce such a topology change, we have an example of a time machine at work. The question becomes whether the laws of classical GTR allow such a change or whether they forbid it. An example where the latter is the case provides an illustration of a CPT, albeit a rather weak one. Suppose that the laws of physics require that spacetime M, g_{ab} admits a *spin structure*. (For a non-compact M this means that M, g_{ab} is *parallelizable*, i.e. there is a C^0, non-vanishing, field of orthonormal tetrads.) This must be the case, for example, if Weyl fermions are to make sense. Then as Hawking and Gibbons (1992) prove, a topology change cannot take place if it consists of the creation of a "wormhole", e.g. the topology goes from $\Sigma = S^3$ in case (a) (or $\Sigma_T = S^3$ in case (b)) to $\Sigma' = S^1 \times S^2$ ($\Sigma'_T = S^1 \times S^2$ in case (b)). This result is interesting but not especially powerful as a CPT since the creation of pairs of wormholes is allowed (see Hawking and Gibbons 1992).

To prove more effective CPTs, additional laws have to be brought into play. EFE require that $R_{ab} - (1/2)Rg_{ab} + \Lambda g_{ab} = 8\pi T_{ab}$, where R_{ab} is the *Ricci tensor*, R is the *curvature scalar*, and Λ is the cosmological constant. The WEC entails the *null energy condition* $T_{ab}K^a K^b \geq 0$ where K^a where K is any null vector. This latter condition and EFE together entail that $R_{ab}K^a K^b \geq 0$, which is called the *null convergence condition*. Finally, the *null generic condition* requires that for every null geodesic there is a point at which the tangent vector K^a satisfies $K^a K^b K_{[c}R_{d]ab[e}K_{f]} = 0$, where R_{abcd} is the Riemann curvature tensor.[9] This condition assures that at some point the geodesic experiences a non-vanishing tidal force. We are now in a position to state a CPT that apparently answers Lemma 1(a) by blocking a topology change in a spatially closed universe.

LEMMA 2 (Tipler 1977). Let M, g_{ab} be a time oriented spacetime satisfying the null convergence and null generic conditions.

(a)   Suppose that $K \subset M$ is a compact set whose boundary is the disjoint union of two compact spacelike three manifolds Σ and Σ'. Then $\Sigma \cong \Sigma'$.

The second part of Lemma 2 uses some additional concepts which require explanation. If Σ is a partial Cauchy surface for M, g_{ab}, the *future domain of dependence* $D^+(\Sigma)$ is defined as the set of all $p \in M$ such that every past directed causal curve which passes through p and which has no past endpoint meets Σ. The *past domain of dependence* $D^-(\Sigma)$ of Σ is defined similarly. Intuitively, $D(\Sigma) = D^+(\Sigma) \cup D^-(\Sigma)$ is the region of spacetime where the state is determined by initial conditions on Σ.[10] Σ is said to be a *Cauchy surface* for M, g_{ab} just in case $D(\Sigma) = M$. If $C \subset \Sigma$ is a connected compact set, Σ is said to be *externally Euclidean* just in case $\Sigma - C$ is diffeomorphically $S^2 \times \mathbb{R}$. If Σ and Σ' are time slices and the portion $T_{\Sigma,\Sigma'}$ of the timelike tube T capped by Σ and Σ' is compact, then the portion of the spacetime $M_B = J^-(\Sigma') \cap J^+(\Sigma)$ between Σ and Σ' is said to be *externally Lorentzian* just in case $M_B - T_{\Sigma,\Sigma'}$ is diffeomorphically $S^2 \times \mathbb{R} \times [0, 1]$, for each $r \in [0, 1]$ $S^2 \times \mathbb{R}$ is spacelike, and for each $x \in S^2 \times \mathbb{R}$ the line $[0, 1]$ is timelike.

LEMMA 2 (continued). As before, let M, g_{ab} be a time oriented spacetime satisfying the null convergence and null generic condition.

(b)   Suppose that M_B is an externally Lorentzian portion of the spacetime between two externally Euclidean time slices Σ and Σ'. If Σ is a partial Cauchy surface for the spacetime and a Cauchy surface for the region $M_B - T_{\Sigma,\Sigma'}$ exterior to the capped tube, then $\Sigma \cong \Sigma'$ and Σ is a Cauchy surface for M_B (i.e. $M_B \subset D^+(\Sigma)$).

Lemma 2(a) shows that the laws of classical GTR preclude the manufacture of CTCs in a generic spatially closed universe by means of a topology change. Unfortunately, this result is effective as a CPT only against one way in which CTCs can develop. Taub-NUT spacetime (Hawking and Ellis 1973, pp. 170–178) shows how a CTC can develop to the future of a compact partial Cauchy surface without any topology change occurring. Figure 1 shows this process for Misner's (1967) two-dimensional spacetime which displays causal features similar to those of Taub-NUT spacetime. Taub-NUT spacetime does not satisfy the null generic condition, so it is not a counterexample to the chronology protection conjecture that under the stated conditions of Lemma 2(a), CTCs do not form in K. As far as I know, this conjecture is open. My own guess is that it fails.

Lemma 2(b) appears to be more effective as a CPT. It says that in a spatially open universe, if potential chronology violations are confined to a compact region $T_{\Sigma,\Sigma'}$ formed by a timelike tube capped by two spacelike hypersurfaces and if determinism holds for the region of spacetime exterior to the tube, then not only does no topology change take place but also

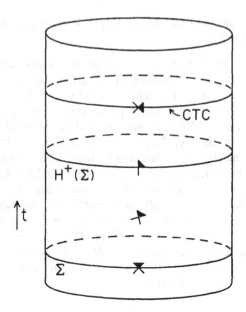

Figure 1. Misner spacetime.

no chronology violation occurs. The latter follows from the fact that $D^+(\Sigma)$ is globally hyperbolic, which implies that it cannot contain closed or almost closed causal curves.[11]

But why, as Lemma 2(b) assumes, should $T_{\Sigma,\Sigma'}$ be compact? The null convergence and null generic conditions are two of the key conditions used to prove the existence of spacetime singularities. And should the singularities be curvature singularities, $T_{\Sigma,\Sigma'}$ cannot be compact. This calls for a little explaining. In the seminal singularity theorems of Hawking and Penrose, 'singularity' stands for geodesic incompleteness.[12] It is possible to have an incomplete null geodesic confined to a compact subset of the spacetime; indeed, Taub-NUT spacetime mentioned above is a relevant example. (In Figure 1 future incomplete null geodesics spiral around the cylinder and remain within a compact neighborhood of the null surface labeled $H^+(\Sigma)$.) However, if the metric is C^2 (as it is in this example) it follows that any scalar curvature polynomial must be bounded so that an incomplete geodesic does not indicate a singularity in the sense of a curvature "blow up". To illustrate mathematically how such a blow up can occur, start with Minkowski spacetime $\mathbb{R}^4$, η_{ab}. Choose a point $p \in \mathbb{R}^4$ and define a scalar field Ω that goes rapidly to 0 as p is approached from any direction. Then for the conformally related metric $g_{ab} = \Omega^2 \eta_{ab}$ the Riemann curvature scalar "blows up" as p is approached. This means that, strictly speaking, the point p is not part of the spacetime with the new metric g_{ab}. The punctured spacetime M, g_{ab}, with $M = \mathbb{R}^4 - p$, is of course geodesically incomplete. The incomplete geodesics are not confined to a compact subset of M since the point p is "missing".

Without further information it is difficult to tell how far curvature singularities may undermine the effectiveness of Lemma 2(b) as a CPT. But all is not lost, for singularities provide another route to proving CPTs. Or so it has been claimed.

3. TIPLER'S CPTS

Tipler (1977) proved a series of mathematical theorems to the effect that if various conditions, which are supposed to characterize the operation of a time machine, hold along with the WEC and the generic condition for null and/or timelike geodesics,[13] then singularities will develop in solutions to EFE. These technical results were interpreted as demonstrating the impossibility of "manufacturing" CTCs. The grounds for this reading were summarized by Tipler:

[A]ny attempt to evolve CTCs from regular initial data will cause singularities to form in the spacetime. Thus, if by the word "manufacture" we mean "construct using only ordinary materials *everywhere*", then the theorems of this paper will conclusively demonstrate that a CTC-containing region cannot be manufactured. For a singularity is a region where the matter density becomes infinite, and matter with arbitrarily large density clearly cannot be considered "ordinary material". (1977, p. 1)[14]

I will not review the details of Tipler's ingenious theorems, but I will challenge the notions that they bear the interpretation Tipler assigns and that, therefore, they deserve to be called CPTs. Note first that the formal theorems demonstrate the existence of singularities in the sense of geodesic incompleteness and do not, without further argument, show that the singularities will be of the strong curvature type that crushes matter down to a zero volume.[15] But even assuming that the singularities are of the strong curvature type, the formal results by themselves constitute CPTs only by virtue of a play on words. Suppose that a would-be time machine is built out of ordinary materials in the sense of steel, plastic, silicon chips, etc. As a result of the operation of the machine, some or all of this material may be crushed to such high density that it no longer qualifies as "ordinary material". But such crushing hardly betokens the kind of chronology protection that was sought unless the stress-energy tensor of the material violates the standard energy conditions. To put it the other way round, the relevant sense of 'ordinary' in "ordinary material" is given by the conjunction of the conditions (i) the material can be assembled from readily available stocks using non-exotic processes and (ii) at no time during the operation does the stress-energy tensor of the matter violate the WEC or other energy conditions.[16]

The point has recently been rubbed in by Ori (1993). Suppose we try to manufacture a black hole by rounding up and transporting to our sun enough nuclear fuel that its mass increases tenfold. According to the standard theory of stellar evolution, the white dwarf core will collapse, resulting in a supernova. The core that is left will continue to collapse to form a neutron star. If this neutron star has a mass greater than 2.5 solar masses, it will be unable to resist a final collapse into a black hole. One will not be impressed if told that one cannot succeed in manufacturing a black hole in this manner because the material employed will not remain "ordinary" when it encounters the black hole singularity.

None of this is to deny that curvature singularities may play a role in a genuine CPT. Such singularities may develop to the future of a partial Cauchy surface Σ in such a way as to prevent any would-be time traveler from starting on Σ and journeying to regions where CTCs exist. If they exist, CTCs must lie on the future side of the future boundary $H^+(\Sigma)$ of $D^+(\Sigma)$ (aka as the *future Cauchy horizon* of Σ).[17] To serve as completely impermeable barriers to all would-be time travelers, the curvature singularities must, intuitively speaking, form all along $H^+(\Sigma)$. Strictly speaking, however, the singularities are not literally in the spacetime. So the envisioned development of singularities will cut off the evolution that started on Σ, with the result that Σ is not just a partial Cauchy surface but a full future Cauchy surface.[18] So the sought after CPT will have the form: Let M, g_{ab}, T^{ab} be a cosmological model satisfying EFE, the WEC, and the generic conditions, and let Σ be a partial Cauchy surface; then if initial conditions on Σ satisfy such and such constraints, Σ is a future Cauchy surface. Such a theorem would not only count as a CPT but would also help to establish Roger Penrose's cosmic censorship hypothesis.[19] Cosmic censorship theorems have been eagerly sought but have proven very difficult to establish. One should, therefore, expect that CPTs of the type envisioned will also prove to be elusive.

4. HAWKING'S CPT

Consider a spacetime M, g_{ab} which admits a partial Cauchy surface Σ. If the future Cauchy horizon $H^+(\Sigma)$ of Σ is not empty, it consists of a null surface whose generators are null geodesics. Hawking says that $H^+(\Sigma)$ is *compactly generated* just in case every past directed null geodesic generator enters and remains within a compact set.

THEOREM (Hawking 1992). Let M, g_{ab}, T^{ab} be a cosmological model, and let $\Sigma \subset M$ be a partial Cauchy surface. Suppose that $H^+(\Sigma)$ is non-empty and is compactly generated. If in addition EFE and the WEC (and, therefore, the null convergence condition) hold, then (a) Σ cannot be non-compact, and (b) whether Σ is compact or non-compact, matter-energy cannot cross $H^+(\Sigma)$.

Part (b) of Hawking's CPT is not especially powerful since (at best) it says that the operator of the time machine cannot be a capitalist because the paying customers cannot take advantage of the fruits of the time machine's operation. Part (a) appears to be much more powerful since reading it in the contrapositive mode seems to imply that in a spatially open universe any attempt to operate a time machine will run afoul of EFE or the WEC. But does the formal mathematical result bear the intended interpretation?

The answer depends largely on the meaning and justification of the crucial condition that $H^+(\Sigma)$ is compactly generated. Physically this condition says that the generators do not "come from infinity" nor do they emerge from a curvature singularity. The first prohibition might seem to be justified as a way of ruling out causality violations that "start at infinity", for such violations cannot legitimately be tied to a time machine, operations of which are supposed to take place in a finite region of space. The second prohibition might seem to be justified as a way of assuring that the development of CTCs is due to the time machine and not to some uncontrollable influence emerging from a singularity. However, a closer look reveals that these motivations are defective.

There is a general problem in saying what it means for the development of CTC's to be "due to" the operation of a time machine or, more generally, to conditions on Σ. "Due to" cannot be cashed in terms of causal determinism. For as already noted, the portion of spacetime where the state is determined by conditions on $\Sigma - D^+(\Sigma)$ – is globally hyperbolic and, therefore, contains no closed causal curves. Furthermore, $H^+(\Sigma)$ is achronal so that chronology violations can take place only to the future of $H^+(\Sigma)$.

One way to try to guarantee that conditions on Σ are responsible for the development of CTCs is to require that these conditions determine that a chronology violation is on the verge of happening in that $H^+(\Sigma)$ contains closed or almost closed null geodesic generators. This requirement will automatically be met if $H^+(\Sigma)$ is compact or compactly generated. (The past directed generators of $H^+(\Sigma)$ for a partial Cauchy Σ are past endless. If they enter and remain within a compact set, then they violate the condition of *strong causality*,[20] which is one precise way of saying that the null generators come arbitrarily close to intersecting themselves.) However, it is not obvious that when $H^+(\Sigma)$ is non-compactly generated the development of CTCs to the future of $H^+(\Sigma)$ is always signaled by closed or almost closed null generators. And in any case, what needs to be assured in order to attribute the development of CTCs to conditions on Σ is that every suitable maximal extension of $D^+(\Sigma)$ contains CTCs. What counts as a "suitable extension" will to some extent vary from case to case. Presumably, we want the extension to be smooth (say C^2). If $D^+(\Sigma)$, g_{ab} is a vacuum solution to EFE, then we will want the extension to also be a vacuum solution. And perhaps we may also want the extension to preserve symmetry properties of the base spacetime. Despite the vagueness, precise results are possible. For example, Taub-

NUT spacetime is a chronology violating vacuum solution to EFE. The Taub portion of the spacetime contains no CTCs. $H^+(\Sigma)$ for a partial Cauchy Σ of this region is compact and is ruled by closed null geodesics. The Taub region admits of different (i.e. non-diffeomorphic) extensions satisfying the vacuum EFE, but all of them contain CTCs. In cases where it is clear that all suitable extensions of $D^+(\Sigma)$ do contain CTCs, then $H^+(\Sigma)$ deserves Hawking's sobriquet of *chronology horizon*.

A further condition is needed to assure that the development of a chronology horizon is traceable back to the manipulation of matter and fields in a finite region of space. Ori (1993) proposed the requirement that $H^+(\Sigma)$ be *compactly causally generated* in that there be a compact $Q \subseteq \Sigma$ such that $H^+(\Sigma) \subset J^+(Q)$. However, this condition is not sufficient to its task since it only guarantees that every point on $H^+(\Sigma)$ can be causally influenced by Q; it leaves open the possibility that there are possibly infinite regions of Σ disjoint from Q that can and do influence $H^+(\Sigma)$. This worry dissolves if *in fact* the conditions in Q are causally responsible for what happens on $H^+(\Sigma)$. As Prof. Ori has remarked,[21] if an accused murderer is charged with shooting his victim to death, it is hardly a defense to admit that the accused fired the fatal shot while noting that it is physically possible that other potentially fatal influences could have but didn't come from regions not under the control of the accused. But it is a contentious matter in general to say what it means to say that X is in fact the cause (or the principal cause) of Y. And in any case a mathematically precise CPT must rely on purely geometrical conditions. Thus, in order to guarantee that the development of the chronology horizon is due to the manipulation of matter and fields in a finite region, I propose that the proper explication of the notion that $H^+(\Sigma)$ is compactly causally generated from Σ is that the topological closure of $I^-(H^+(\Sigma)) \cap \Sigma$ – which is the portion of Σ that can causally influence $H^+(\Sigma)$ – is compact.

A few examples will help to illustrate what is involved. In the Misner spacetime pictures in Figure 1, $I^-(H^+(\Sigma)) \cap \Sigma = \Sigma$, and Σ is compact. So $H^+(\Sigma)$ is compactly generated in Hawking's sense and is compactly generated in both Ori's and my sense. In the case of Reissner–Nordström spacetime (which does not contain CTCs) whose Penrose conformal diagram is given in Figure 2, it is also true that $I^-(H^+(\Sigma)) \cap \Sigma = \Sigma$; but since Σ is non-compact, $H^+(\Sigma)$ is not compactly generated in Hawking's sense nor is it compactly generated in my sense; it is, however, compactly causally generated in Ori's sense. An artificial example of a compactly causally generated $H^+(\Sigma)$ for a non-compact Σ is given in Figure 3. This example also illustrates how $H^+(\Sigma)$ can be compactly causally generated in my sense without being compactly generated in Hawking's sense.

Requiring that all of $H^+(\Sigma)$ be compactly causally generated may be overkill. To weaken the requirement, call $V \cap H^+(\Sigma)$ a *chronology violation inducing* subset just in case any suitable extension of $I^-(V) \cap D^+(\Sigma)$ contains CTCs.[22] It seems sufficient for the operation of

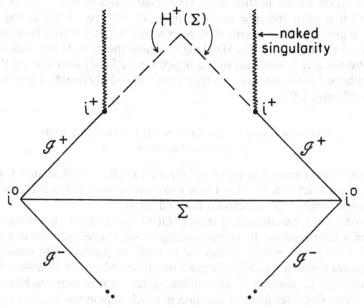

Figure 2. Conformal diagram of a portion of Reissner–Nordström spacetime.

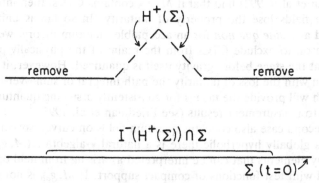

Figure 3. Two-dimensional Minkowski spacetime with two half-lines removed.

a time machine that there be such a V which is compactly causally generated from Σ, i.e. the closure of $I^-(V) \subset \Sigma$ is compact.

My suggestion is that an effective CPT along the lines suggested by Hawking would have to substitute for Hawking's condition that $H^+(\Sigma)$ is compactly generated the condition that there is a chronology violation inducing $V \subset H^+(\Sigma)$ which is compactly causally generated from Σ. For in that case even if the chronology horizon is not compactly generated in Hawking's sense – because generators of $H^+(\Sigma)$ emerge from singularities – there is a clear sense in which the development of CTCs is due to

15

conditions on a finite portion of Σ. That singularities as well as CTCs are created is a price the time machine operator may be willing to pay in order to give his customers the thrill of experiencing time travel. However, the proof techniques used in Hawking's original theorem do not suffice in any obvious way to establish such a hoped for CPT. My own guess is that such a hoped for theorem is not true unless so heavily qualified as not to be an effective CPT.

5. IMPLICATIONS OF QUANTUM FIELD THEORY AND QUANTUM GRAVITY

The above discussion was set within classical GTR. In this section I will make a few brief remarks about how quantum considerations may add to or change some of the conclusion reached above.

On one hand, quantum field theory (QFT) may facilitate the construction of a time machine by countenancing exotic matter that violates the WEC. Using such matter it may be possible to produce and maintain wormholes through which CTCs can thread (see Morris et al. 1988). On the other hand, quantum considerations militate against time machines in three ways. Consider first the situation in which a quantum field is put on a fixed relativistic spacetime M, g_{ab} without any attempt to quantize gravity or even to calculate the backreaction of the field on the metric g_{ab}. Friedman et al. (1992) find that if M, g_{ab} contains CTCs, then interacting quantum fields lose the property of unitarity. In so far as unitarity is regarded as a *sine qua non* for an acceptable quantum theory, we have a solid reason to exclude CTCs from the realm of the physically possible, at least at the stage before gravity itself is quantized. However, it may be that even with the loss of unitarity the path integral or sum-over-histories approach will provide the means for consistently assigning quantum probabilities to measurement results (see Friedman et al. 1992).

The second case also concerns quantum fields on curved spacetimes. If M, g_{ab} is globally hyperbolic there is a natural $*$-algebra $A(M, g_{ab})$ generated by elements that can be interpreted as arising from field operators smeared with test functions of compact support. If M, g_{ab} is not globally hyperbolic there may or may not be a natural $A(M, g_{ab})$. But Kay (1992) has argued that whatever the choice of field algebra $A(M, g_{ab})$ it should satisfy the requirement of *F-locality*: every neighborhood of each point of M should contain a subneighborhood N such that the restriction of the global algebra should coincide with the natural algebra $A(N, g_{ab}|_N)$ of the globally hyperbolic spacetime $N, g_{ab}|_N$. If there is no $A(M, g_{ab})$ satisfying F-locality, Kay says that the spacetime is *non-quantum compatible*. He proves that the Misner spacetime, which evolves CTCs from a non-chronology violating region, is non-quantum compatible. It could be conjectured that any cosmological model that satisfies EFE and that illustrates the operation of a time machine is non-quantum compatible.

16

The third case concerns semi-classical quantum gravity where the metric itself is not quantized but the quantum expectation value $\langle T^{ab} \rangle$ of the (renormalized) stress-energy tensor is calculated and the backreaction on the metric is computed by inserting $\langle T^{ab} \rangle$ back into EFE. In some concrete cases of spacetimes where CTCs develop to the future of a partial Cauchy surface Σ, it has been found that $\langle T^{ab} \rangle$ diverges as the chronology horizon $H^+(\Sigma)$ is approached. If the divergence is strong enough, it can cut off future development and thus prevent the formation of CTCs. This matter is under discussion in the physics literature (see Boulware 1991; Hawking 1992; Kim and Thorne 1991; and Klinkhammer 1992).

6. CONCLUSIONS

Quantum considerations may wreck dreams of operating a time machine and may even prove to be inconsistent with any version of time travel, whether induced by a time machine or not. However, this matter can only be settled when we have in hand a viable theory of quantum gravity, and such a theory continues to elude our grasp. It seems prudent, therefore, to try to understand as best we can the implications of classical GTR for the prospects of operating a time machine. This task turns out to be more difficult than one might have anticipated, in part because it is not easy to carry out the prior task of elucidating the conditions that are essential for characterizing a time machine in the setting of classical GTR. In this paper I have made an effort to carry out this prior task. If I have been successful, then I think it follows that none of the CPTs that have been offered in the setting of classical GTR succeed in showing that time machines are beyond the pale of the physically possible. I have suggested some forms that a CPT would have to take in order to succeed in placing time machines beyond this pale. And I have suggested also that it is unlikely that such theorems will be proven in the foreseeable future, either because they are not theorems or else because they are exceedingly difficult to prove. Such theorems, if they exist, would have independent interest because not only would they succeed in outlawing time machines but would also demonstrate versions of cosmic censorship.[23]

NOTES

[1] It is assumed throughout that the spacetimes under discussion are time orientable so that a globally consistent time directionality can be assigned. How the future direction is chosen is part of the problem of the direction of time, a problem that will not be broached here. The signature of the spacetime metric g_{ab} is chosen to be $(+ + + -)$.

[2] Let V^a be the normalized $(V^a V_a = -1)$ four-velocity of γ. The four-acceleration vector is $A^a = V^b \nabla_b V^a$, where ∇_a is the covariant derivative operator associated with g_{ab}. The magnitude of acceleration is $a = (A^a A_a)^{1/2}$. Thus, the total acceleration $TA(\gamma)$ along a curve γ is $\int_\gamma a \, d\tau$, where τ is proper time along γ.

[3] T^{ab} describes the distribution of matter-energy in spacetime. EFE will be stated in Section 2.

[4] The grandfather paradox arises from the observation that if time travel were possible, then it would also seem to be possible for a person to travel into her own distant past and shoot her grandfather at a stage of his life before he sired any children, leading to an absurdity. In Earman (1995) I argue that these paradoxes are simply crude ways of bringing out the fact that in spacetimes with CTCs local initial conditions are constrained in unfamiliar ways.

[5] There are also cosmological models which satisfy EFE with zero cosmological constant and which contain CTCs.

[6] This is the solution of van Stockum; see Tipler (1974).

[7] For $X \subset M$, $J^-(X)$ stands for all points $p \in M$ such that there is a future directed causal curve from p to X. Similarly, the causal future of $J^+(X)$ of X is the set of all $q \in M$ such that there is a future directed causal curve from X to q.

[8] If 'time machine' is meant in the weak sense (see Section 1), then many of the science fiction stories about time machines can be accomodated in relativistic spacetime models.

[9] The square brackets denote anti-symmetrization of indices.

[10] This assumes that the field equations which govern gravitational phenomena are of the hyperbolic type so that existence and uniqueness of solutions corresponding to initial data on Σ can be proved.

[11] The formal definition of global hyperbolicity is given in Hawking and Ellis (1973, pp. 206ff). A spacetime is globally hyperbolic if and only if it possesses a Cauchy surface.

[12] These singularity theorems are reviewed in Hawking and Ellis (1973). A *half geodesic* is a geodesic that has one end point and is extended as far as possible in some direction from that point. Such a geodesic is said to be *incomplete* just in case its affine length is finite. For timelike and spacelike geodesics this means that the proper length is finite.

[13] The generic condition for timelike geodesics is similar in formulation to the generic condition for timelike geodesics; it implies that a timelike geodesic will, at some point, experience a tidal force.

[14] I have substituted CTC for Tipler's CTL (closed timelike line).

[15] Tipler (1985) has given a formal criterion for such crushing singularities.

[16] The *strong* and *dominant energy conditions* impose additional constraints on T^{ab}; see Hawking and Ellis (1973, pp. 91, 95).

[17] $H^+(\Sigma) = D^+(\Sigma) - I^-(D^+(\Sigma))$. Here $I^-(X)$, $X \subset M$, stands for the *chronological past* of X, i.e. the set of all $p \in M$ such that there is a non-trivial future directed timelike curve from p to X.

[18] That is, $J^+(\Sigma) \subseteq D^+(\Sigma)$.

[19] For a review of literature on this subject, see Earman (1993).

[20] See Lemma 8.21 of Wald (1984). The *strong causality condition* holds for a spacetime M, g_{ab} just in case for every point $p \in M$ and any open neighborhood $N(p)$ of p, there is a subneighborhood $N'(p)$ such that once a causal curve exits $N'(p)$ it never reenters.

[21] Private communication.

[22] I am indebted to C. J. S. Clarke for this suggestion.

[23] I am grateful to Michael Friedman and John Norton for helpful suggestions on an earlier draft of this paper.

REFERENCES

Boulware, D. G.: 1992, 'Quantum Field Theory in Spaces with Closed Timelike Curves', *Physical Review D* **46**, 4421–4441.

Earman, J.: 1993, 'The Cosmic Censorship Hypothesis', in J. Earman, A. I. Janis, G. Massey, and N. Rescher (eds.), *Philosophical Problems of the Internal and External Worlds: Essays On the Philosophy of Adolf Grünbaum*, University of Pittsburgh Press, Pittsburgh, Pennsylvania.

Earman, J.: 1995, 'Recent Work on Time Travel', in S. Savitt (ed.), *Time's Arrow Today*, Cambridge University Press, Cambridge, in press.

Friedman, J. L., Papastantiou, N. J., and Simon, J. L.: 1992, 'Failure of Unitarity for Interacting Fields in Spacetimes with Closed Timelike Curves', *Physical Review D* **46**, 4456–4476.

Geroch, R. P.: 1967, 'Topology in General Relativity', *Journal of Mathematical Physics* **8**, 782–786.

Hawking, S. W.: 1992, 'Chronology Protection Conjecture', *Physical Review D* **46**, 603–611.

Hawking, S. W. and Ellis, G. F. R.: 1973, *The Large Scale Structure of Space-Time*, Cambridge University Press, Cambridge.

Hawking, S. W. and Gibbons, G. W.: 1992, 'Selection Rules for Topology Change', *Communications in Mathematical Physics* **148**, 345–352.

Kay, B. S.: 1992, 'The Principle of Locality and Quantum Field Theory on (Non Globally Hyperbolic) Spacetimes', *Reviews of Mathematical Physics*, Special Issue, 167–195.

Kim, S.-W. and Thorne, K. S.: 1991. 'Do Vacuum Fluctuations Prevent the Creation of Closed Timelike Curves?', *Physical Review D* **43**, 3929–3947.

Klinkhammer, G.: 1992, 'Vacuum Polarization of Scalar and Spinor Fields near Closed Null Geodesics', *Physical Review D* **46**, 3388–3394.

Malament, D.: 1985, 'Minimal Acceleration Requirements for 'Time Travel' in Gödel Spacetime', *Journal of Mathematical Physics* **26**, 774–777.

Misner, C. W: 1967, 'Taub-NUT Space as a Counterexample to Almost Everything', in J. Ehlers (ed.), *Relativity Theory and Astrophysics I: Relativity and Cosmology*, American Mathematical Society, Providence, Rhode Island, pp. 160–169.

Morris, M. S., Thorne, K. S., and Yurtsever, U.: 1988, 'Wormholes, Time Machines, and the Weak Energy Condition', *Physical Review Letters* **61**, 1446–1449.

Ori, A.: 1993, 'Must Time-Machine Construction Violate the Weak Energy Condition?', *Physical Review Letters* **71**, 2517–2520.

Tipler, F. J.: 1974, 'Rotating Cylinders and the Possibility of Global Causality Violation', *Physical Review D* **9**, 2203–2206.

Tipler, F. J.: 1977, 'Singularities and Causality Violation', *Annals of Physics* **108**, 1–36.

Tipler, F. J.: 1985, 'Note on Cosmic Censorship', *General Relativity and Gravitation* **17**, 499–507.

Wald, R.: 1984, *General Relativity*, University of Chicago Press, Chicago.

Dept. of History and Philosophy of Science
University of Pittsburgh
Pittsburgh PA 15620
U.S.A.

ROBERT ALAN COLEMAN AND HERBERT KORTÉ

A NEW SEMANTICS FOR THE
EPISTEMOLOGY OF GEOMETRY I:
MODELING SPACETIME STRUCTURE

1. INTRODUCTION

In this paper, we present a new semantics for the epistemology of geometry which provides the necessary foundation for modeling spacetime structure, in particular for modeling the physical concepts and procedures used for the empirical determination of geometric and geometric-object fields, by introducing a number of crucial concepts and distinctions such as:

(1)   formal, theoretic and physical coordinates;

These coordinate concepts and distinctions permit a deeper understanding of covariance and invariance, in particular of the fact that physical descriptors are invariants under formal active and/or passive transformations, but they are covariants under a transformation from one physical coordinate system to another physical coordinate system.

(2)   symmetry versus model diffeomorphisms;

This distinction and the previous distinction are necessary for a coherent formulation of the measurement problem in GTR; moreover, they are again required for a coherent formulation of Einstein's 'Hole' (Loch) problem, a formulation which leads to the result that causality in GTR is strictly deterministic independently of whether a field relationalist, field-body relationalist or a manifold substantivalist ontology of spacetime is adopted.

(3)   the active versus the passive use of technology;

The use of technology solely for the passive purpose of assigning physical coordinates to events together with a coordinate independent analysis of tracking data is the necessary prerequisite for constructing a noncircular and nonconventional discovery procedure for the empirical determination of the physical descriptors of geometric and geometric-object fields.

(4)   theoretic completeness, epistemic completeness, and decidability with respect to the post-differential-topological, geometric structures of a spacetime theory;

These notions are crucial to a clear conception of the constructive approach to spacetime theories.[1]

(5)   the concept that a geometric field is a G-structure on spacetime.

Because a G-structure is a field of frames and/or coframes and because frames and coframes are by definition nondegenerate, a geometric field

Erkenntnis **42**: 141–160, 1995.

21

can never vanish anywhere in contrast to for example an electromagnetic field (a geometric-object field). This insight leads to the realization that the field-body relationalist view of the structure of the world constitutes a coherent ontology. There are no unoccupied points.

As briefly indicated above, the semantic analysis of the epistemology of geometry presented in this paper bears on three important topics in the philosophy of spacetime: the conventionalist-realist debate, the relationalist-substantivalist debate and the debate concerning the nature of causality in GTR, Einstein's 'Hole' problem. In this paper and in the paper [2, Coleman and Korté, appearing in this issue] we confine the application of our semantic analysis to the conventionalist-realist debate, that is, to the problem of showing how the physical geometry of spacetime can be determined empirically in a nonconventional manner. For extensive discussion of the application of our semantic analysis to the relationalist-substantivalist debate and to the debate concerning causality in GTR, the reader is referred to our previous work [10, 12].

The organization of the paper is as follows. In the section on formal, theoretic and physical coordinates, we first discuss the nature of these types of coordinates and the roles they play, then we specify what is involved in the modeling of physical coordinates and then we present concrete models of physical coordinates in STR and GTR. In the next section, a number of principles pertaining to the nature of measurement of geometric fields are presented. Then, theoretic completeness, epistemic completeness and epistemic decidability with respect to the geometric structures of a theory are defined and discussed. Finally, we make some historical remarks pertaining to theoretic completeness of Einstein–Maxwell theory.

In the paper [2, Coleman and Korté, appearing in this issue], we apply our semantic analysis presented here and show that Galilean theory is epistemically complete with respect to its spatial metric and that Einstein–Maxwell theory is epistemically complete with respect to its spacetime geometry.

2. MATHEMATICAL, THEORETICAL AND PHYSICAL COORDINATES

2.1. *Physical Coordinates*

The manner in which a physical theory is formulated, constructed and presented corresponds, roughly speaking, to a kind of modeling process of the physical world. The purpose of this section is to analyze aspects of this process, namely the nature, type and roles of coordinates within this modeling process, and to sketch the evolution of the coordinate concept from pre-GTR to GTR. Within the context of GTR, the nature and functions of coordinates within the modeling process are both subtle and complex. Indeed, it is through the difficulties that arise in the case of

GTR that one becomes aware of the fact that certain steps of the modeling process, that cannot be ignored in the case of GTR, are in fact also necessary for the formulation of pre-GTR theories. Even though a naive approach to the process of assigning coordinates to events in the world is possible in the context of pre-GTR theories, an understanding of this process, its coherence and consistency requires a theoretical analysis of the physical objects or processes used for the construction of a physical coordinate system. Even in the case of a coordinate system based on the use of rigid rods, for example, a theoretical account of such coordinates requires at least the proof (a standard result of classical mechanics) that the concept of 'rigid rod' is a coherent notion in classical mechanics (that is, rigid rods can exist within classical mechanics). It will become clear in the light of the analysis presented below that these steps could be skipped over only with the aid of assumptions that now appear ad hoc.

Let us first consider Galilean models of spacetime. Such spacetime theories presuppose a variety of geometric structures that are both absolute and flat. In particular, space is assumed to be Euclidean; moreover, the spatial geometry is typically introduced theoretically by *stipulating* the existence of a system of spatial coordinates (x, y, z) with respect to which the spatial metric is given by

$$ds^2 = dx^2 + dy^2 + dz^2. \tag{2.1}$$

At this theoretical stage of the presentation, no physical apparatus for setting up physical coordinate systems has been introduced and the coordinates (x, y, z) are to be regarded as non-physical coordinates. On the other hand, these coordinates are not merely *formal* or *purely mathematical* coordinates, that is coordinates which have no specified relation to any post-differential-topological, theoretical and/or physical structures postulated by the theory. They have not been formally stipulated by the theorist in an arbitrary manner for the sole purpose of labeling the points of space in a differential-topologically consistent manner. Rather the coordinates (x, y, z) are determined up to a Euclidean transformation by a fundamental element of the theory, the spatial metric, and hence are linked to the metrical structure postulated by the theory. Moreover, the metric structure postulated by the theory represents something physical in the world and for its physical determination appropriate procedures (for example, the use of physical coordinates) may be introduced at a later stage of the modeling process. For this reason, we call the coordinates (x, y, z) introduced in the above manner **theoretic** coordinates.

It is easy to overlook the *theoretic* character of the coordinates (x, y, z) in pre-GTR theories because intuition suggests the immediate introduction of rigid rods which one may use to construct physical coordinate systems that are **adapted** to the physical geometry. Moreover, to each of the *theoretic* coordinate systems adapted to the spatial metric of the theory given by (2.1), there corresponds a physical coordinate system – set up by physically constructing a grid of rigid rods – that is adapted to the *physical* spatial metric. In a similar manner, the *theoretic* time coordinate with respect to which the absolute time structure of the world is described in the Galilean model of spacetime is readily provided with a physical counterpart by the introduction of ideal clocks. Two things should be noted. First, the physical objects which have been introduced in the modeling process, namely rigid rods and ideal clocks, have a complex structure. Second, they are being used in this context in an **active** as opposed to a purely **passive** manner since their role is a double one: not only are they being used **passively** to set up a physical coordinate system by providing a physical means of **merely** labeling points arbitrarily in a manner consistent with the differential topology, but they are at the same time used in an **active** way; that is, it is assumed that the spatial and temporal metrics are specified functions of these physical coordinates. In particular, the spatial metric is assumed to be given by (2.1) with respect to the physical coordinates set up using rigid rods. In other words, rigid rods (and ideal clocks) are also used to determine the **spatial metric** and the **time metric**. Clearly, if rigid rods and ideal clocks are used in this active *or metric* way, in setting up physical coordinate systems, then the metrics that are thereby determined are mere conventional stipulations, since the question whether or not a physical rod qualifies as a *rigid* rod cannot be answered in an independent non-circular way. Rigid rods and ideal clocks have been introduced into the modeling process not merely to serve in a passive capacity, but to play an active role by providing a physical basis for fundamental post-differential-topological structures, *without any prior theoretical analysis of these physical objects*. The use of rigid rods and ideal clocks to set up physical coordinates which are adapted to this geometry makes use of this geometry implicitly.[2]

In the case of the Special Theory of Relativity (STR), an absolute, flat, spacetime metric is presupposed; moreover, it is assumed that a system of coordinates (t, x, y, z) exists with respect to which the spacetime metric is given by

$$ds^2 = -dt^2 + d\vec{r} \cdot d\vec{r}. \tag{2.2}$$

Again, the coordinates (t, x, y, z) are *theoretic*. They are determined up to a Poincaré transformation by the fact that they are adapted to a fundamental element of the theory, the spacetime metric. Once again, the fact that such coordinates are *theoretic* and not physical was in large measure skipped over. Physical counterparts to the coordinates (t, x, y, z) were introduced early in the presentation of the theory by employing quasi-rigid rods and quasi-ideal clocks to construct a *physical* coordinate system or frame of reference; moreover, the quasi-rigid rods and quasi-ideal clocks were used actively; that is, it was stipulated that in terms of the physical coordinates, the spacetime metric was given by (2.2).[3]

An important feature shared by these two cases is the existence of a

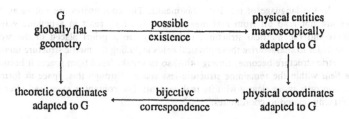

Fig. 1. Physical coordinates in pre-GTR theories.

globally *flat* geometry *G* which makes possible the existence of physical entities, such as rods and clocks which are macroscopically (not just at a point) adapted[4] to the geometry *G*. This circumstance leads to a bijective correspondence between the set of theoretic coordinate systems that are adapted to the metric structure *G* postulated by the theory and the set of physical coordinate systems that are adapted to the physical geometry represented by *G*. Figure 1 summarizes the situation.

In the case of Einstein–Maxwell theory (roughly speaking GTR together with electromagnetism), the situation is much more complicated and subtle. The spacetime metric is not an absolute structure with convenient global symmetries; rather, it is a dynamical physical entity that is coupled to the energy-momentum density of matter and other fields. In almost all cases, the spacetime metric does not have any symmetries at all.[5] The theory tells us that the best we can do is to adapt to a microneighbourhood by employing a normal coordinate system at a given point. In the absence of symmetry, extended bodies adapted to the geometry cannot exist. In such circumstances, physical coordinate systems cannot be introduced early in the presentation of the theory even with the aid of ad hoc assumptions because the circumstances that permitted the ad hoc assumptions in the pre-GTR case simply do not obtain. A proper theoretic account of the epistemology of geometry requires an analysis of the physical measurement process itself, in particular an analysis of physical coordinate systems. In Einstein–Maxwell theory, the description of physical coordinate systems is an advanced topic in the sense that many other physical entities, including the spacetime metric, the electromagnetic field, the world lines of material bodies and their equation-of-motion structures, have to be introduced and analyzed first. Clearly, the description and analysis of these physical entities must be carried out with respect to a purely mathematical or formal system of coordinates because physical coordinates are not yet part of the model. As Weyl puts it in the following quote, it is necessary to employ coordinates that are **not measured**.

Coordinates are introduced on the Mf [Manigfaltigkeit (manifold)] in the most direct way through the mapping onto the number space, in such a way, that all coordinates, which arise through one-to-one continuous transformations, are equally possible. With this the coordinate concept breaks lose from all special constructions to which it was bound earlier

25

in geometry. In the language of relativity this means: The coordinates are not *measured*, their values are not read off from real measuring rods which react in a definite way to physical fields and the metrical structure, rather they are a priori placed in the world arbitrarily, in order to characterize those physical fields including the metric structure numerically. The metric structure becomes through this, so to speak, freed from space; it becomes an existing *field* within the remaining structure-less space. Through this, space as form of appearance contrasts more clearly with its real content: The content is measured after the form is arbitrarily related to coordinates.[6]

Even in the case of Einstein–Maxwell theory, physical coordinates make use of the geometry even if not in the uniform manner that rigid rods and ideal clocks do. The signals emitted by a radar station are, for example, guided by the conformal structure of spacetime. In contrast with the geometric structures of classical mechanics, the geometric structure of Einstein–Maxwell theory is not given a priori, rather it is part of the dynamical problem and hence physical coordinates are also part of the dynamical problem. The relation between the physical coordinates and the geometric structure in Einstein–Maxwell theory is described by a complicated system of functions $g_{ij}(x^i)$. The corresponding relationship in classical mechanics is described by a small number of constants; for example, $g_{\alpha\beta}(x^\alpha) = \delta_{\alpha\beta}$ for the spatial metric.

In the case of pre-GTR, Galilean theories, the use of purely mathematical, formal coordinates is not necessary because of the assumptions pertaining to the *flatness* of the geometric structures which permits the use of theoretic coordinates that are adapted to the geometry; nevertheless, a proper theoretical account of the epistemology of geometry still requires an analysis of the physical measurement process including an analysis of physical coordinates with respect to these theoretic coordinates. For the case of the spatial metric, such an analysis involves proving that the concept of a rigid rod is consistent with Newtonian mechanics,[7] and the provision of a sequence of filters (properties that rigid rods must possess) each of which is, however, testable at the level of the differential topology, that is, by observing coincidences. We show how this can be done below in [2, Coleman and Korté, appearing in this issue]. These theoretic analyses are carried out with respect to the theoretic coordinates. It was the fact that physicists made an ad hoc assumption regarding the metric character of rigid rods (their *active* employment in the measurement process) rather than developing appropriate differential topological criteria on the basis of a theoretical analysis, that led to the assertion by conventionalists such as Reichenbach that the spatial metric was conventional.

2.2. *Theoretical Modeling of Physical Coordinate Charts*

Ideally, a spacetime theory should provide a complete analysis of the physical coordinate systems with respect to which the physical descriptors of its physical fields, its physical geometric fields in particular, are to be determined. In practice, however, this goal cannot be achieved because any physical coordinate chart will make use of some devices that cannot be completely analyzed within the theory because the ontological domain of the theory is limited in scope.

For definiteness, consider the case of Einstein–Maxwell theory. We regard the ontological domain of Einstein–Maxwell theory to include, roughly speaking, such entities and/or processes as:

(1) The gravitational interactions and dynamics of macroscopic matter distributions.
(2) The electromagnetic interactions and dynamics of macroscopic charge distributions.
(3) The propagation of electromagnetic fields and their interaction with macroscopic charge distributions.

On the other hand, the large number of phenomena that are accounted for on the basis of quantized or semi-quantized versions of electromagnetism lie strictly outside of the ontological domain of Einstein–Maxwell theory.

We have described elsewhere [9] a radar-station coordinate chart that is based on electromagnetic signals emitted and received by a single observer. This type of chart requires the determination of emission and reception times of electromagnetic signals as well as the direction (θ, φ) of the incoming signals. An instantiation of a physical radar-station chart may be found at any large airport. We also present below a physical radar chart that employs three observers. This system is somewhat simpler in that only emission and reception times of electromagnetic signals need be determined.

The important features of these models, the motion(s) of the base station(s), the propagation of the electromagnetic signals and the essentially elementary (pointlike) nature of the emission, scattering and reception events, can be accounted for within Einstein–Maxwell theory; moreover, the procedure for assigning coordinates to the scattering event can also be justified within the theory. The clock (possibly an atomic clock) used to assign times to the emission and reception events and the device used to determine the values of (θ, φ) in the case of a radar-station chart, however, lie strictly outside the ontological domain of Einstein–Maxwell theory. These models of physical coordinates may, nevertheless, be considered reasonable for the following reasons:

(1) The fact that the devices that are not strictly within the ontological domain of Einstein–Maxwell theory are used *solely* for the assignment of physical coordinates to scattering events and that subsequent analysis of the tracking data for the measurement process is coordinate independent means that the devices are being used **actively** only with respect to the differential topology of spacetime,

but are being used strictly **passively** with respect to post-differential-topological structures, in particular, with respect to geometric structures. It is only assumed that the devices behave smoothly with respect to the differential topology.

(2) There exist theoretical parameters, t and (θ, φ) to which the numerical outputs of these devices correspond in an appropriate manner, respecting, for example, the monotonically increasing character of t and the finite bounds on (θ, φ).

In summary, what is required is a theoretical justification for the usual pragmatic presupposition; namely, that when one makes use of a physical coordinate system in practice, that is, when one operates in the world with certain pieces of physical equipment and uses the numerical outputs obtained to assign physical coordinates to physical events in the world (physical spacetime), one presupposes that the procedures followed do in fact yield quadruples of real numbers that are smoothly related to the physical differential topology of spacetime:

(1) We observe that the differential topology of a spacetime theory is presented theoretically by means of an atlas of local, purely mathematical or formal coordinate charts.

(2) The epistemology of geometry is concerned with providing a theoretical justification (as complete as possible) of the empirical procedures used, not just with providing a pragmatic justification.

(3) To give a theoretical account of a physical coordinate chart within a given theory, one must describe the **relevant** physical processes that give rise to the physical coordinates. This description must be with respect to a purely mathematical or formal coordinate system for otherwise the analysis would be circular.

(4) The analysis must then show that if the ontological presuppositions of the theory are true, then the physical coordinates that would be assigned to an event that is designated by given formal coordinates are related by a passive transformation from the formal to the physical coordinates, a pfP-transformation.

2.3. *Physical Coordinates in STR and GTR*

Let us first consider the case of a relatively simple system of physical coordinates in STR. It is postulated that there exists a flat pseudo-Riemannian metric. As a consequence of a mathematical theorem, we know that there exist systems of adapted, theoretic coordinates x^i_t in the sense that the metric in the interval is everywhere given by

$$d\underset{t}{s}^2 = -d\underset{t}{t}^2 + d\underset{t}{\vec{r}} \cdot d\underset{t}{\vec{r}}. \tag{2.3}$$

These coordinates are not physical in the sense that they are some

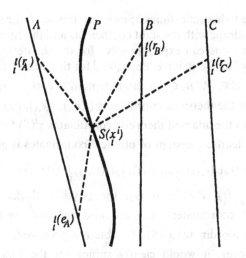

Fig. 2. Physical coordinates in STR.

functions of numerical outputs of some actual **physical** mensuration devices. Rather, they are coordinates that are *supposed* to have a special relationship to a fundamental entity (the spacetime metric) that is postulated by the theory. The physical entities and processes that result in the assignment of physical coordinates $x_p(S)$ to an event S designated by the theoretic coordinates $x_t^i(S)$ are described with respect to the theoretic coordinates x_t^i as follows:

(1) The base of the coordinate system consists of three *inertial* observers (A, B, C) with world paths (unparameterized curves) described by

$$x_t^\alpha = \xi_t^\alpha(t)$$

$$= \xi_t^\alpha(t_0) + \xi_t^\alpha(t_0)(t - t_0), \quad Z \in \{A, B, C\}. \qquad (2.4)$$

(2) The propagation of light is governed by the spacetime metric (2.3) given above.

(3) Each of the observers A, B, C is equipped with a clock of some sort (light clock, atomic clock). The output of each of these clocks is given by a function of t.

$$t_p A(t), t_p B(t), t_p C(t). \qquad (2.5)$$

which we shall initially assume to be linear. The situation is illustrated in Figure 2.

The particle P is present in order to provide an object from which the signal emitted from A at the event e_A can scatter in order to propagate along the forward light cone at S to reach the reception events r_A, r_B and

29

r_C at the indicated theoretic times. Since for this case the lightcone is described by a quadratic with constant coefficients and the equations of the paths are all linear, one can explicitly solve for the theoretic coordinates $(\xi_A^\alpha(\underset{t}{t}(e_A)), \underset{t}{t}(e_A))$ of the emission event e_A and for the theoretic coordinates $(\xi_Z^\alpha(\underset{t}{t}(r_Z)), \underset{t}{t}(r_Z))$, $Z \in \{A, B, C\}$ of the reception events r_A, r_B and r_C. It can be proved that the theoretic coordinates $(\underset{t}{t}(e_A), \underset{t}{t}(r_A), \underset{t}{t}(r_B), \underset{t}{t}(r_C))$ are smoothly related to the adapted theoretic coordinates $\underset{t}{x^i}(S)$ for some open neighbourhood. Clearly, a system of physical coordinates is given by

$$\underset{p}{x^i}(S) = (\underset{p}{t_A}(\underset{t}{t}(e_A)), \underset{p}{t_A}(\underset{t}{t}(r_A)), \underset{p}{t_B}(\underset{t}{t}(r_B)), \underset{p}{t_C}(\underset{t}{t}(r_C))). \tag{2.6}$$

If the functions $\underset{p}{t_Z}(\underset{t}{t})$, $Z \in \{A, B, C\}$, the physical clocks, are linear, then the physical coordinates $\underset{p}{x^i}(S)$ are also smoothly related to the adapted theoretic coordinates $\underset{t}{x^i}(S)$. To obtain a good system of physical coordinates, however, it would clearly suffice for the functions $\underset{p}{t_Z}(\underset{t}{t})$, $Z \in \{A, B, C\}$ to be diffeomorphisms; consequently, one could use atomic clocks for this purpose of assigning physical coordinates, provided only that atomic time is *smoothly* (differential-topologically) related to gravitational time but not necessarily metrically related. In addition, one need not require that the observers A, B and C are inertial; however, it would then be more difficult to carry out the computations to determine the domain in which the coordinates would be valid.

The only remaining generalization is the transition to the case of Einstein–Maxwell theory in which the spacetime metric is deformed so that the curvature is nonzero. Then, there no longer exist theoretic coordinates $\underset{t}{x^i}$ with respect to which the components of the metric are simple constants (η_{ij}). Theoretic coordinates that are functions of the curvature invariants [15, Kretschmann, 1917] or that are due to partial symmetries of the metric (Killing vector fields) exist. In fact the theoretic coordinates $\underset{t}{x^i}$ used above in the flat case are Killing coordinates. We will discuss Kretschmann–Killing theoretic coordinates elsewhere. To determine Kretschmann–Killing theoretic coordinates that are appropriate for a spacetime metric, *one must know* that metric; consequently, in any general analysis of the concept of physical coordinates in the Einstein–Maxwell case, it is necessary, initially at least, to abandon theoretic coordinates altogether and resort instead to the use of *formal*, purely mathematical[8] coordinates because the physical metric is simply not known.

(1) The three world line paths are described by

$$\underset{f}{x^\alpha} = \underset{f}{\xi_Z^\alpha}(\underset{f}{t}), \quad Z \in \{A, B, C\} \tag{2.7}$$

with respect to the formal coordinates $\underset{f}{x^i}$.

(2) The spacetime metric is described by

30

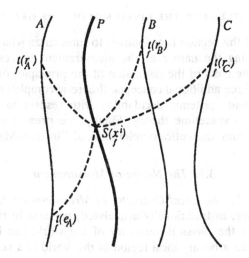

Fig. 3. Physical coordinates in Einstein–Maxwell theory.

$$g = \underset{f}{g}_{ij}(\underset{f}{x}^i)\, d\underset{f}{x}^i \otimes d\underset{f}{x}^j.$$

(3) The physical clocks on A, B, C are supposed to correspond to three diffeomorphisms

$$\underset{p}{t}_Z = T_Z(\underset{f}{t}_Z), \quad Z \in \{A, B, C\}. \tag{2.9}$$

From the formally described equation for the three paths A, B, C and from the spacetime metric, one can in principle determine the formal coordinates of the emission event e_A and of the reception events r_A, r_B and r_C corresponding to the event S designated by the formal coordinates $\underset{f}{x}^i(S)$. Then from the diffeomorphisms T_Z, $Z \in \{A, B, C\}$ that describe the behaviour of the physical clocks in terms of the formal time along the paths, one can determine the physical coordinates for S, namely

$$\underset{p}{x}^i(S) = (\underset{p}{t}_A(\underset{f}{t}(e_A)), \underset{p}{t}_A(\underset{f}{t}(r_A)), \underset{p}{t}_B(\underset{f}{t}(r_B)), \underset{p}{t}_C(\underset{f}{t}(r_C))). \tag{2.10}$$

One requirement for theoretic completeness of Einstein–Maxwell theory is that one has to show that the physical coordinates $\underset{p}{x}^i(S)$ are diffeomorphically related to the formal coordinates $\underset{f}{x}^i(S)$ in some formally described open neighbourhood provided that the functions that describe the paths A, B, C and the spacetime metric g with respect to the formal coordinates $\underset{f}{x}^i$ are suitably restricted and the functions that describe the physical clocks T_Z, $Z \in \{A, B, C\}$ are diffeomorphisms. The functions T_Z, $Z \in \{A, B, C\}$ should not be further restricted in order to permit, for example, the use of atomic clocks.

3. PRINCIPLES FOR THE EPISTEMOLOGY OF GEOMETRY

The purpose of this section is as follows: to summarize what has been said so far concerning the nature of the measurement process. Second, to introduce on the basis of the discussion of the principles of the measurement process three additional concepts: theoretic completeness, epistemic completeness and epistemic decidability with respect to the geometric structure(s) of a spacetime theory. Finally, we present some historical remarks concerning theoretic completeness of Einstein–Maxwell theory.

3.1. *The Nature of Measurement*

PRINCIPLE 3.1. *The Local Character of Measurements*: The epistemology of geometry, as traditionally articulated, pertains to the question of whether or not the physical geometry of the world can be empirically determined in an arbitrary local region of the world in a nonconventional manner. The issue of the conventionality of geometry should not be conflated with the epistemology of cosmology.

Geometric-object fields[9] are locally described by a system of components (descriptors) with respect to a local coordinate chart. In the context of pure mathematics the local coordinates are merely stipulated or assumed; that is, their existence is not in any way physically real. We shall call such purely mathematical coordinates which do not in any way refer to the real world, **formal** coordinates and call the corresponding components **formal** descriptors. On the other hand, the measurement of a *physical* geometric-object field is the empirical determination of its **physical** descriptors, that is, its system of components with respect to a local **physical** coordinate chart.

PRINCIPLE 3.2. *The Nature of Geometric Measurements*: The measurement of a physical geometric-object field is the empirical determination of its local descriptors with respect to a local, physical coordinate chart.

As we have discussed elsewhere [10, 12], physical *descriptors* are covariant with respect to a change from one physical coordinate system to another physical coordinate system, that is, under a ppP-transformation (physical physical passive). The physical situation can, of course, be described with respect to various local, *formal* coordinate charts. We emphasize that under a change from one *formal* coordinate chart to another *formal* coordinate chart, that is, under an ffP-transformation (formal formal passive), the *physical* descriptors of the geometric-object field are *invariants*.

Clearly, it is necessary to employ a local, physical coordinate system for the measurement of a physical geometric field such as the spacetime metric; moreover, such coordinate systems, for example, a radar-station chart, typically are rather complex entities and their construction makes use of advanced technology based on sophisticated physical theories including the theory of geometry that is the subject of investigation. These coordinate systems are in fact designed to be as adapted as possible to what is thought to be the geometry of the world; that is, every effort is made to make the coordinates as convenient and as 'nice' as possible.

Superficially, it would appear that using such a coordinate system for the measurement of geometric fields would involve a vicious circularity; however, as we noted above there is **no** circularity at all provided that the technology is used **passively**, that is, strictly for the assignment of coordinates to events or locations, and provided that the criteria and analyses used in the measurement process are coordinate independent.

PRINCIPLE 3.3. *The Passive Use of Technology*: In the direct measurement of physical entities such as fields and particle paths, the use of the procedures and products of advanced technology is permissible provided that they are used only passively, that is, solely for the physical assignment of coordinates to scattering events (for the tracking of particles and light signals), and provided that the criteria and mathematical analyses applied to the data are coordinate independent.

The very notion of a test object or of a probe that is used for the measurement of a physical structure, the existence of which is postulated by a theory, is of necessity theory-laden because the theory tells us what physical objects, should they exist, would be suitable probes, and the theory provides criteria regarding their behaviour under relevant circumstances. In the case of classical mechanics, for example, the existence of ideal clocks and rigid rods is possible, but their existence is not guaranteed theoretically a priori because the ideal behaviour of such complex objects could only be established theoretically by a very complicated analysis of the dynamics of matter. In other words, the theory essentially tells us only that the concepts of an ideal clock or rigid rod make sense.

In the case of GTR, the theory provides for the possible existence of neutral massive monopoles and tells us that the motions of these particles would be governed by the projective structure of spacetime and that their motions would reveal this structure. In addition, the Einstein–Maxwell theory also postulates the existence of electromagnetic waves, that the propagation of these waves is governed by the conformal structure of spacetime and that these waves would be suitable probes for measuring the conformal structure.

PRINCIPLE 3.4. *Theory Dependence of Probes Used for Measurement*: The probes and procedures that are suitable for the measurement of a physical geometric structure are specified by the very theory that postulates the existence of that structure.

The above mentioned theory dependence of probes and procedures used for the measurement of geometric structures does not entail that theory independent, pre-geometric, coordinate independent criteria do not exist

(1) for the physical identification of the probes should they exist
(2) for their behaviour under relevant test conditions and
(3) for the collection and coordinate independent analyses of the experimental data used to determine the physical descriptors of the geometric field with respect to a local *physical* coordinate chart.

Indeed, we shall discuss in [2, Coleman and Korté, appearing in this issue] two spacetime theories, namely, Newton–Galilei theory and Einstein–Maxwell theory, and show that in the context of both theories it is possible to formulate the requisite criteria in purely differential topological and coordinate independent terms; that is, we show that Newton–Galilei theory and Einstein–Maxwell theory both satisfy the following epistemological desideratum:

DESIDERATUM 3.5. *Theory Independent, Pre-Geometric Measurement Criteria*: The criteria derived from the postulates of a theory for the identification of probes, for the behaviour of these probes and for the collection and analyses of the data used in the measurement of a geometric field are testable at the level of differential topology.

3.2. *Degrees of Completeness and Decidability of Spacetime Theories*

The above analysis leads to the following definition of *theoretical completeness* of a spacetime theory.

DEFINITION 3.6. *Theoretical Completeness*: A spacetime theory is theoretically complete with respect to its post-differential topological geometric fields, if and only if the theory provides a model of a measurement procedure for the geometric fields which the theory postulates exist, that is, if and only if

(1) **Physical Coordinates:** The theory provides a top-down theoretical analysis as to how physical entitles and/or processes that are within the ontological domain of the theory could be used to set up physical coordinates that are (theoretically) compatible with the local differential topology.
(2) **Probes:** The theory provides a theoretical analysis that shows how the postulated behaviour of suitable entities (probes) and/or pro-

cesses that belong to the ontological domain of the theory could in principle be used to obtain the physical descriptors of the geometric fields postulated by the theory with respect to the physical coordinate systems.

A spacetime theory which can merely be shown to be theoretically complete with respect to its post-differential-topological geometric fields does not provide a noncircular empirical discovery procedure for the empirical determination of the post-differential-topological geometric fields (such as, a metric field) which it postulates. Such a theory is therefore vulnerable to a conventionalist critique. To escape the conventionalist challenge one must be able to show that a spacetime theory, besides being theoretically complete, is also *epistemologically complete*.

DEFINITION 3.7. *Epistemological Completeness*: A theory is epistemically complete with respect to its geometric fields if and only if

(1) The theory is theoretically complete and
(2) The theory provides additional differential-topological analyses and/or criteria which are in principle testable at the level of the local differential topology in a coordinate independent manner for
    (a) the identification of the probes (should they exist) to be used in the measurement process,
    (b) the determination that the probes behave in the manner postulated by the theory and
    (c) the determination of the physical descriptors of the geometric fields with respect to the physical system of coordinates.

Note the qualification 'should they exist' in item (2)(a) of the Definition 3.7. Epistemological completeness merely asserts that suitable probes, should they exist, will provide us with epistemic access to post-differential-topological geometric fields postulated by the theory, since it can be shown that there exist local differential-topological and coordinate independent criteria for their identification, for the testing of their behaviour and for the determination of the physical descriptors of the geometric fields with respect to a physical system of coordinates.

The post-differential-topological geometric fields are **revealed** by, and not defined by the probes. They are not ontologically reducible to nor do they depend relationally on the actual presence of suitable probes. For example, optics and electrodynamics played essentially only a heuristic role in the construction of Special Relativity. The latter's validity or justification does not depend in an essential way on the actual existence of electromagnetic waves. Properly understood, the velocity of light is an invariant upper limiting velocity. Special relativistic mechanics says only that there is an upper limiting velocity. It does not say that there exist entities that propagate with that velocity. The conformal-causal structure postulated by Special Relativity is not ontologically reducible to nor does it depend relationally on the *actual* existence of electromagnetic waves. The existence of light is a gift of nature which certainly aided in the construction of Special Relativity but it is not necessary for the validity of the theory. Spacetime geometry is not about probes such as rigid rods, ideal clocks, electromagnetic waves or freely falling particles, except in the derivative sense of providing information about the physically real post-differential-topological geometric fields.

DEFINITION 3.8. *Epistemic Decidability*: A spacetime theory is epistemically decidable with respect to its post-differential-topological geometric fields if and only if

(1) the theory is epistemically complete and
(2) the probes mentioned in item (2)(a) of the Definition 3.7 of epistemological completeness actually exist.

It should be clear that a spacetime theory which can be shown to be epistemologically complete with respect to the post-differential-topological geometric fields which it postulates is *nonconventional* with respect to that structure. The circularity charges typically levelled by conventionalists against empirical discovery procedures for the post-differential-topological geometric fields implicitly exploit the fact that any description of the behaviour of actual or hypothetical probes which does not make exclusive use of the local differential topology, must avail itself of additional geometric structures. Since such post-differential-topological structures are alleged to be conventional – either on verificationist (Reichenbach) or ontological (Grünbaum) grounds – different post-differential topological geometric hypotheses may conventionally be chosen that will lead to different *geometric* descriptions of the same set of observable properties and relations realized by the hypothetical or actual probative systems. Any putative discovery procedure will therefore be unsuccessful if one requires for a characterization of the observable behaviour of the probes used, geometric knowledge beyond the local differential topology. For example, the causal-inertial method for the discovery of the spacetime metric would be circular, if it were the case that one required for the geometric characterization of light propagation and particle motion those very local post-differential-topological structures, namely, the conformal causal and projective structures, which are supposed to be revealed through their behaviour. The possibility of a noncircular and hence nonconventional discovery procedure of the post-differential-topological geometric fields requires that the theory be epistemologically complete.

3.3. *Theoretical Completeness of Einstein–Maxwell Theory*

In this section, we briefly sketch the historical development that led to the proof that Einstein–Maxwell theory is theoretically complete. That Einstein–Maxwell theory is also epistemically complete was established much later by the development of a measurement procedure for the projective structure of spacetime [3, 4, 5, 6, 7, 8, 11].

In his 1917 paper [15], Kretschmann discussed three important topics. Only the first, his treatment of general covariance, has been generally recognized and discussed in the recent literature. He also discussed theoretic coordinates as we have noted above. The third item constituted the first theoretical analysis which showed that if two spacetime metrics have the same light rays and the same free fall paths, then those metrics are the same up to a spacetime independent constant. He first showed that if the metrics have the same light rays, then the spacetime metrics are the same up to a (possibly) spacetime dependent scale factor. From the constraint associated with the free-fall paths, he showed that the scale factor is determined up to a factor that is independent of the spacetime location; moreover, although Kretschmann formulated the constraint regarding the free-fall paths in terms of the equation-of-motion for the geodesics of a symmetric linear connection, he was essentially dealing with the paths rather than the curves because the parameters of the curves were eliminated in the calculation. The analysis was carried out using what we have called formal coordinates.

In a letter to Felix Klein published in 1921 [19], Weyl presented a result similar to that of Kretschmann. Weyl abstracted the conformal structure from the spacetime metric and the projective structure from the symmetric linear connection, and he proved that the conformal and projective structures that correspond to a given spacetime metric together uniquely determine that metric. Weyl's result is somewhat clearer in that it identifies precisely the spacetime structures that govern the propagation of electromagnetic waves and the motions of neutral massive monopoles; that is, Weyl's explication of the relationship between the dynamics of the physical probes and the geometric structures of spacetime is explicit.

Two years later Lorentz [16, Lorentz, 1923] presented in a more explicit manner Kretschmann's procedure for determining the spacetime metric. Lorentz presented a thought experiment in which large numbers of material particles and light rays in a region of spacetime collide to produce a *rich* fabric of coincidences which are identified with the physically observable facts.[10] Lorentz used a system of purely formal coordinates, constrained only by the requirement that the relations between the coincidences are smoothly represented, to describe the pattern of physical events. Focusing on a particular spacetime event, Lorentz noted that *nine* light rays through that event determined the metric coefficients at that event up to a constant that could depend on the coordinates of that event.

He also noted that each additional light ray through this event provided a *test* of this determination. Lorentz also used the equation for the *geodesic* through the event to obtain an equation for the scale factor along that path; moreover, *he explicitly noted that he eliminated the affine parameter along the geodesic* thereby making explicit what was only implicit in Kretschmann's presentation. In this way he showed that light propagation and free-fall motion could be used to determine the metric of spacetime.

Much later (1972), Ehlers, Pirani and Schild [13] presented a Constructive Axiomatics for GTR (that is, Einstein–Maxwell theory). They improve the result of Weyl in that they consider conformal and projective structures on spacetime that need not a priori be related to or derived from a spacetime metric. Among other things, they derive a condition that must be satisfied if these two structures are compatible in the sense that the solution paths of the projective structure (free-fall paths) cannot *break the light barrier*. They point out that if this condition is satisfied, the two structures determine a Weyl structure which reduces to a spacetime metric structure provided that there are no second clock effects (the Streckenkrümmung vanishes). In addition, EPS address the other aspect of what we have called theoretic completeness, namely, a theoretical analysis of physical coordinate systems. They describe a system of physical radar coordinates similar to the physical radar coordinates presented above; moreover, they point out in a footnote that in the case of STR, it is a theorem that such coordinates are smoothly related to the (what we call) theoretic coordinates that are adapted to the flat spacetime metric at least in some open neighbourhood. It would appear that they regarded this as a justification of the use of physical radar coordinates in the case of GTR because locally the transition from STR to GTR involves a smooth local deformation of the fields and particle paths that determine the physical coordinates.

NOTES

* This work was carried out while the authors were participants in a research group at the Zentrum für interdisziplinäre Forschung at the Universität in Bielefeld, Germany, during the academic year 1992–1993.
[1] Contributions to the constructive approach may be found in [1, 3, 4, 5, 6, 7, 8, 11, 13, 15, 17, 18, 19, 22].
[2] It also makes use of the notion that such objects can be dynamically isolated from other bodies and/or fields in the world.
[3] Of course the prefix 'quasi' is necessary in the case of STR because of the known effects called Lorentz contraction and time dilation. Because of these effects, however, it does not even make sense to use such physical entities actively to stipulate the spacetime metric until appropriate criteria for inertiality and mutuality of rest in addition to an analysis of the behaviour of rods and clocks have been provided. Even if such criteria were to exist, however, it is still not appropriate to make metrical assumptions about the associated physical coordinates, that is, to assume that the spacetime metric is given by (2.2) with respect to these coordinates.

[4] Of course, other cases might exist for which the geometry is not flat but has constant curvature and for which other physical procedures permit the construction of globally adapted physical coordinate systems. The consideration of such possibilities would not, however, add anything of importance to our narrative.

[5] Of course, most of the known exact solutions of the field equations exhibit considerable symmetry precisely because such solutions are easier to find.

[6] Hermann Weyl [21].

[7] In particular, one needs the standard result [14, See Goldstein, p. 10] that the constraints $r_{ij} = c_{ij}$, where r_{ij} is the distance between particles i and j of the body and the c_{ij} are constants, entail that the internal potential of the body is a constant provided that the internal forces are radial. The kinematics and dynamics of the body can then be treated in terms of its center-of-mass motion and its rotational motion about its center of mass. Only the consistency of the kinematic notion of a rigid body is required in this context. No claim as to the actual existence of a suitable system of forces that would guarantee that rigid bodies exist is made.

[8] In the case of flat spacetimes (Minkowskian or Galilean), one technically needs to employ formal, purely mathematical coordinates as well, but only to the extent that one should first prove that there exist theoretic coordinates that are adapted to the flat structures postulated by these theories before using the theoretic coordinates in further analyses.

[9] The usage of the terms 'geometric-object field' and 'geometric field' is such that the first includes the second. By a 'geometric field', we mean a G-structure that corresponds to some aspect of the post-differential-topological geometry of a manifold such as a Riemannian, conformal, affine or projective structure. Besides these structures, the term 'geometric-object field' also includes such fields as equation-of-motion structures for massive monopoles and the electromagnetic field. See Appendix A.2 of [2, Coleman and Korté, appearing in this issue] for a brief characterization of an $O(\mathbb{R}^n)$-structure ($G = O(\mathbb{R}^n)$).

[10] Incidentally, this identification of coincidences with that which is physically observable is reminiscent of Einstein's approach to the resolution of his 'Hole' problem which Einstein first sketched in a letter to Ehrenfest.

REFERENCES

[1] Castagnino, M.: 1971, 'The Riemannian Structure of Space-Time as a Consequence of a Measurement Method', *Journal of Mathematical Physics* 12, 2203–2211.

[2] Coleman, R. A. and Korté, H.: 1995, 'A New Semantics for the Epistemology of Geometry II: Epistemological Completeness of Newton–Galilei and Einstein–Maxwell Theory', this issue.

[3] Coleman, R. A. and Korté, H.: 1980, 'Jet Bundles and Path Structures', *The Journal of Mathematical Physics* 21(6), 1340–1351.

[4] Coleman, R. A. and Korté, H.: 1982, 'The Status and Meaning of the Laws of Inertia', in *The Proceedings of the Biennial Meeting of the Philosophy of Science Association*, Philadelphia, pp. 257–274.

[5] Coleman, R. A. and Korté, H.: 1984, 'Constraints on the Nature of Inertial Motion Arising from the Universality of Free Fall and the Conformal Causal Structure of Spacetime', *The Journal of Mathematical Physics* 25(12), 3513–3526.

[6] Coleman, R. A. and Korté, H.: 1987, 'Any Physical, Monopole, Equation-of-Motion Structure Uniquely Determines a Projective Inertial Structure and an $(n - 1)$-Force', *The Journal of Mathematical Physics* 28(7), 1492–1498.

[7] Coleman, R. A. and Korté, H.: 1989, 'All Directing Fields that are Polynomial in the $(n - 1)$-Velocity are Geodesic', *The Journal of Mathematical Physics* 30(5), 1030–1033.

[8] Coleman, R. A. and Korté, H.: 1990, 'Harmonic Analysis of Directing Fields', *The Journal of Mathematical Physics* 31(1), 127–130.

[9] Coleman, R. A. and Korté, H.: 1992, 'On Attempts to Rescue the Conventionality Thesis of Distant Simultaneity in STR', *Foundations of Physics Letters* **5**(6), 535–571.

[10] Coleman, R. A. and Korté, H.: 1991, 'The Relation between the Measurement and Cauchy Problems of GTR', in H. Sato and T. Nakamura (eds), *The Sixth Marcel Grossmann Meeting on General Relativity*, pp. 97–119. World Scientific, 1992. Printed version of an invited talk presented at the meeting held in Kyoto, Japan, 23–29 June 1991.

[11] Coleman, R. A. and Korte, H.: 1994, 'Constructive Realism', in U. Majer and H.-J. Schmidt, (eds), *Semantical Aspects of Spacetime Theories*, Wissenschaftsverlag, Mannheim u.a., pp. 67–81.

[12] Coleman, R. A. and Korté, H.: 1994, 'A Semantic Analysis of Model and Symmetry Diffeomorphisms in Modern Spacetime Theories, in U. Majer and H.-J. Schmidt (eds), *Semantical Aspects of Spacetime Theories*, Wissenschaftsverlag, Mannfeim u.a., pp. 83–94.

[13] Ehlers, J., Pirani, R. A. E., and Schild, A.: 1972, 'The Geometry of Free Fall and Light Propagation', in L. O. Raifeartaigh (ed.), *General Relativity*, *Papers in Honour of J. L. Synge*, Clarendon Press, Oxford, pp. 63–84.

[14] Goldstein, H.: 1950, *Classical Mechanics*, Addison-Wesley Publisher Company, Inc., Reading, Massachusetts. Sixth printing.

[15] Kretschmann, E.: 1917, 'Über den physikalischen Sinn der Relativitätstheorie', *Annalen der Physik* **53**(16), 576–614.

[16] Lorentz, H. A.: 1923, 'The Determination of the Potentials in the General Theory of Relativity, with Some Remarks about the Measurement of Length and Intervals of Time and about the Theories of Weyl and Eddington', *Proc. Acad. Amsterdam* **29**, 363–382.

[17] Pirani, F. A. E.: 'Building Space-Time from Light Rays and Free Particles', *Symposia Mathematica* **XII**, 67–83.

[18] Reichenbach, H.: 1969, *Axiomatization of the Theory of Relativity*, University of California Press, Los Angeles.

[19] Weyl, H.: 1921, Zur Infinitesimalgeometrie: Einordnung der projektiven und konformen Auffassung. *Nachr. Königl. Ges. Wiss. Göttingen*, *Math.-phys. Kl.*, Reprinted in [20], pp. 99–112.

[20] Weyl, H.: 1968, *Gesammelte Abhandlungen*, volume 1–4. Springer Verlag, Berlin, edited by K. Chandrasekharan.

[21] Weyl, H.: 1988, *Riemanns geometrische Ideean, ihre Auswirkung und ihre Verknüpfung mit der Gruppentheorie*, Springer-Verlag, Berlin. Edited by K. Chandrasekharan.

[22] Woodhouse, N.: 1973, 'The Differentiable and Causal Structures of Space-Time', *Journal of Mathematical Physics* **14**, 495–501.

Received 8 November 1994

University of Regina
Regina
Saskatchewan
Canada S4S 0A2

ROBERT ALAN COLEMAN AND HERBERT KORTÉ

A NEW SEMANTICS FOR THE EPISTEMOLOGY OF GEOMETRY II:
EPISTEMOLOGICAL COMPLETENESS OF NEWTON-GALILEI AND EINSTEIN-MAXWELL THEORY

1. INTRODUCTION

In [1, Coleman and Korté, appearing in this issue] we present the principles and analyses underlying a new approach to the epistemology of spacetime geometry. In this paper we apply this analysis to two spacetime theories. In particular, we show that Newton–Galilei theory is epistemically complete with respect to its spatial metric and that Einstein–Maxwell theory is epistemically complete with respect to its spacetime geometry.

The proof that Newton–Galilei theory is both theoretically and epistemically complete with respect to the spatial metric postulated by that theory is presented in the next section. This proof makes use of the Lie-pseudo-group-structure description of geometric structure. This approach to the description of geometric structure is applicable only in the case that the structure admits global symmetries. For the case of a Riemannian structure, this means that one is dealing with a manifold with constant positive curvature, constant zero curvature or constant negative curvature. The case of importance for this paper is that of constant zero curvature, that is, Euclidean geometry. A simple proof of the fact that the Lie-pseudo-group and G-structure descriptions of Euclidean geometry are equivalent is presented in Section A.2 of the appendix for the convenience of the reader. This result appears in the mathematical literature [19, 20, 24] but in an extremely general context.

In Section 3 of this paper, we outline the proof that Einstein–Maxwell theory is both theoretically and epistemologically complete. In this case of course, the geometry does not in general have any global symmetries; consequently, the analysis is presented in terms of the more familiar G-structure description of pseudo-Riemannian geometry. We reformulate and organize our previous results [2, 5, 6, 8, 9, 10, 13, 14, 15, 16] in terms of the new semantics for the epistemology of geometry that we present in [1, Coleman and Korté, appearing in this issue].

41

Erkenntnis **42**: 161–189, 1995.

2. A NONCONVENTIONAL DISCOVERY PROCEDURE FOR THE EUCLIDEAN SPATIAL METRIC IN NEWTON–GALILEI THEORY

2.1. *Theoretical Completeness of Newton–Galilei Theory with Respect to its Spatial Metric*

Entities called *rigid bodies* are part of the ontological domain of Newton–Galilei theory. As noted in the footnote 8 of [1, Coleman and Korté, appearing in this issue], the concept of a rigid rod is a coherent notion within Newtonian mechanics. In addition, it is understood within the context of this theory that entities called rigid rods have metrical properties with respect to the Euclidean spatial metric postulated by the theory. The following theorems of 2-dimensional and 3-dimensional Euclidean geometry

THEOREM 2.1. The equilateral paving of the Euclidean plane is rigid.

THEOREM 2.2. The regular octahedral-tetrahedral paving of 3-dimensional Euclidean space is rigid.

provide the basis for physical coordinate systems, discussed in detail below, which are related in a simple way to orthonormal coordinates with respect to which the spatial metric is given by either

$$(2.1) \quad ds^2 = dx^2 + dy^2$$

or by

$$(2.2) \quad ds^2 = dx^2 + dy^2 + dz^2$$

respectively. The theory, is therefore, theoretically complete with respect to the spatial metric which it postulates because it provides a top-down theoretical analysis that shows how physical entities that are within the ontological domain of the theory could in principle (that is, in theory) be used to set up physical coordinates and to determine the physical descriptors of the spatial metric with respect to these physical coordinates. This level of analysis, however, does not entail that the theory is epistemically complete with respect to its spatial metric because no theoretical analysis that provides pre-geometric differential-topological criteria for the identification of and employment of rigid rods has been supplied. From the point of view presented in this paper and in the paper [1, Coleman and Korté, appearing in this issue] it is essentially the absence of this aspect of the analysis which led conventionalists to charge that the use of rigid rods for the empirical determination of the spatial metric was beset with vicious circularity.

2.2. *The Differential-Topological Criteria for the Identification and Employment of Rigid Rods*

As noted above, it is understood within the context of Newton–Galilei theory that entities called rigid rods are endowed with metrical properties; however, the direct employment of these 'metric' entities for the empirical determination of the spatial metric postulated by the theory, does not constitute a non-circular discovery procedure of the spatial metric because the relevant class of rods is simply stipulated to remain congruent under transport and such a stipulation cannot itself be verified without further stipulations of a similar nature. We shall show that this circularity can be avoided.[1] To do so, it is necessary to provide a sequence of tests or filters designed to discover a class of rigid rods of a given length such that each test can be verified at the level of the local differential topology, that is, by observing a sufficient number of coincidences. These filters determine nested sequences of increasingly smaller equivalence classes of rods. Only if a set of rods passes *all* of the tests are the rods rigid and define an equivalence class of rigid rods of a given length. There is an equivalence class of rigid rods corresponding to each distinct length.

The first equivalence relation is that of local congruence defined as follows:

DEFINITION 2.3. *Local Congruence.* Consider two rigid rods with end-points A, B and C, D. Suppose that at some spatial location and for some spatial orientation, it is possible for AB and CD to be mutually at rest with A in coincidence with C and B in coincidence with D simultaneously. One says that AB and CD are congruent at that spatial location for that spatial orientation.

In addition, the theory tells us that rigid bodies have at least the following property:

PROPOSITION 2.4. *Homogeneity and Isotropy of Local Congruence.* Any two rigid rods AB and CD that are congruent for some spatial location and some spatial orientation are congruent for every spatial location and every spatial orientation.

REMARK 2.5. We shall call rods that satisfy Definition 2.3 and Proposition 2.4, HILC-rods.[2] Note that the above two criteria that the theory tells us HILC-rods must satisfy are testable at the level of the local differential topology, because one need only observe a sufficient number of spatial coincidences.

The following propositions and constructions are motivated by Theorems 2.1 and 2.2 in Section 2.1. Newton–Galilei theory informs us that rigid rods satisfy the following proposition.

PROPOSITION 2.6. *HILC-Paving Condition*. Pavings similar to those mentioned in Theorems 2.1 and 2.2, but which need not necessarily be regular, may be satisfied by subclasses of HILC-rods. The rods of such a subclass will be called PCHILC-rods, where the initial 'PC' stands for 'paving closure'.

REMARK 2.7. Once again it is worth emphasizing that to determine whether or not a paving with physical rods succeeds requires only the observation of a sufficient number of spatial coincidences, a criterion which is purely differential topological in character. The pavings for the two-dimensional and three-dimensional cases are illustrated in Figure 1. and Figure 2.

PROPOSITION 2.8. *The Partitioning Rule*. For each of some equivalence classes of PCHILC-rods, there exists an equivalence class of PCHILC-rods that are 'half the length' of the rods in the original class in the sense that

(1)   in the 2-dimensional case, a triangular region consisting of 4 small triangles having 9 of the smaller rods as sides is such that each side is congruent with any side of a triangle formed from 3 of the larger rods.

(2)   in the 3-dimensional case, a tetrahedral region consisting of 4 small tetrahedra and one octahedron having 24 of the smaller rods as edges is such that each face is congruent with each face of a tetrahedron formed from 6 of the larger rods.

In either case the procedure can be iterated indefinitely.

REMARK 2.9. Note that each rod of the larger figure, referred to in Proposition 2.8, corresponds to the composite of two of the smaller rods in the corresponding composite figure; however, it is not asserted that the common point of the composite rod is the midpoint of the longer rod nor does the proposition depend on a claim that the pavings are regular, that is, at this point, the possibility of smooth coherent deformations is not ruled out.

2.2.1. *The Two-Dimensional Case*

In the 2-dimensional case, the grid is completely determined by the choice of an origin ○ and the choice of a spatial direction in which the rod is pointed. In addition, the grid can be refined to any desired degree in the neighbourhood of any point by making use of the rule of partition stated

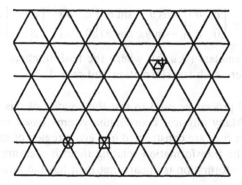

Figure 1. Physical coordinates from a triangular paving.

in Proposition 2.8. While keeping in mind the Remark 2.9, we shall nevertheless refer to the partitions obtained as half-steps, quarter-steps and so forth. Since these units are used **only** for the purpose of assigning coordinates and **not** for the purpose of assigning distances, **no** circularity arises.

With regard to Figure 1, the physical coordinates (α, β) of the point marked with '+' with respect to the origin 'O' and the point '□' with coordinates $(1, 0)$ are approximately

(2.3) $\alpha = 1 + \frac{1}{2} + \frac{1}{4} + \frac{1}{8}$

$\beta = 2 + \frac{1}{2} - \frac{1}{4} + \frac{1}{8}$.

As we have noted above in Section 3 of [1, Coleman and Korté, appearing in this issue], we regard the epistemology of geometry to be a local problem. Focus on some open neighbourhood (portion) of space and suppose that a number of physical pavings of the kind discussed above have been constructed in this region. Let (α, β) and $(\bar{\alpha}, \bar{\beta})$ denote the physical coordinates associated with two of these grids. One can then *empirically* determine the coordinate transformation

(2.4) $\bar{\alpha} = \bar{A}(\alpha, \beta)$
$\bar{\beta} = \bar{B}(\alpha, \beta)$

and its inverse. Newton–Galilei theory informs us that

PROPOSITION 2.10. *Euclideanness.* There exists exactly one one-parameter[3] family of equivalence classes such that, for any two physical coordinate charts (α, β) and $(\bar{\alpha}, \bar{\beta})$ constructed in the manner presented above (with a given PCHILC-class chosen as the standard unit), the coordinate transformation (A, B) and its inverse are similar to Euclidean transformations; that is, the transformation (A, B) has the form

$$(2.5) \quad \begin{bmatrix} \bar{\alpha} \\ \bar{\beta} \end{bmatrix} = \begin{bmatrix} c & d \\ 0 & e \end{bmatrix} \begin{bmatrix} \cos\theta & \sin\theta \\ -\sin\theta & \cos\theta \end{bmatrix} \begin{bmatrix} c & d \\ 0 & e \end{bmatrix}^{-1} \begin{bmatrix} \alpha \\ \beta \end{bmatrix} + \begin{bmatrix} c & d \\ 0 & e \end{bmatrix} \begin{bmatrix} a \\ b \end{bmatrix}$$

for suitable parameters (θ, a, b), where the real numbers c, d and e are the same for every such transformation and $ce > 0$.

The empirical procedure used to test the validity of this proposition is the following. Many physical coordinate systems based on a particular PCHILC-class of rods are constructed in a given region of space. For each pair of charts, the transformation relating them is measured. It will *turn out* that this transformation is linear; that is, it has the form

$$(2.6) \quad \begin{bmatrix} \bar{\alpha} \\ \bar{\beta} \end{bmatrix} = Z \begin{bmatrix} \alpha \\ \beta \end{bmatrix} + \begin{bmatrix} \zeta \\ \eta \end{bmatrix},$$

where the elements of the matrix Z and ζ, η are empirically determined. One obtains in this way a collection $\{Z\}$ of such empirically determined matrices. It is claimed that there exists a single upper triangular, invertible matrix S such that each element of the transformed collection $\{S^{-1}ZS\}$ is a rotation; that is, for each element, there is a rotation R such that $S^{-1}ZS = R$. Since rotations satisfy the constraint $R^T R = I$, each of the elements yields a constraint on the three components of S, namely,

$$(2.7) \quad S^T Z^T S^{-1T} S^{-1} ZS = I,$$

where the elements of Z are empirically known. One can use various analytical techniques, such as the methods of least squares and goodness of fit, to determine the best values for the elements of S and the degree of confidence or goodness of fit of the hypothesis to the data. If this coordinate independent analysis of the data succeeds, then one can conclude that there exists an Euclidean Lie pseudogroup structure or an $\mathbb{E}(\mathbb{R}^2)$-structure on the given portion of space. Equivalently, there exists a flat Euclidean metric on the given portion of space; moreover, the physical coordinates determined by

$$(2.8) \quad \begin{bmatrix} x \\ y \end{bmatrix} = S^{-1} \begin{bmatrix} \alpha \\ \beta \end{bmatrix} = \begin{bmatrix} \alpha + \beta/2 \\ (\sqrt{3})\beta/2 \end{bmatrix}$$

are adapted to the 2-dimensional spatial metric which is given by

$$(2.9) \quad ds^2 = dx^2 + dy^2$$

with respect to the physical coordinates (x, y). The mathematical justification for this last claim is presented below in Section 2.3.

2.2.2. *The Three-Dimensional Case*

For the 3-dimensional case, the steps down to the end of Remark 2.9 are

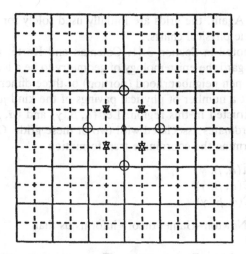

Figure 2. Coordinates from an octahedral-tetrahedral paving.

the same as for the 2-dimensional case except for the indicated differences. In the 3-dimensional case, however, the PCHILC-rods are used to pave the 3-dimensional space in an octahedral-tetrahedral pattern that is not assumed a priori to be regular. A regular octahedral-tetrahedral paving is illustrated in Figure 2. In this figure, two planar square grids are indicated in solid and dashed lines respectively. These grids are parallel to each other and alternate but are separated in space in such a way that each vertex is the same distance from its four nearest neighbours in the two adjacent grids as it is from its four nearest neighbours in its own grid. The three-dimensional grid in a region is a stack of such planar grids alternating from one type to the other. In the figure, the twelve equidistant neighbours of the vertex marked $\diamond$ are indicated, four in the same plane marked $\bigcirc$, four in the plane above marked $\triangle$ and four in the plane below marked ∇. For a mathematical analysis see E. Crapo [17] and W. Whiteley [27].

Here again, in the construction of a physical paving the problematic term 'equidistant' is replaced by the constructive fact that vertices that are nearest neighbours are connected by a HILC-rod. Just as in the two-dimensional case, whether or not the construction succeeds can be empirically verified by observing a sufficient number of coincidences. Clearly, several such physical grids could be constructed in a given region U of the 3-dimensional space; moreover, associated in a straight-forward manner with each such grid is a physical coordinate chart $(U, (\alpha, \beta, \gamma))$ where the oblique coordinates α, β, γ are obtained by counting steps, half-steps, quarter-steps and so forth from the origin $\diamond$ along the obvious lattice directions, in the plane of one of the square grids for α and β and upward along the direction that joins the origin to one of the vertices in an adjacent plane for γ preferably chosen so that (α, β, γ) form a right-handed system. Again, the grids are initially used solely for the purpose of establishing the physical charts.

As we have noted in Section 3 [1, Coleman and Korté, appearing in this issue], we regard the epistemology of geometry to be a local problem. Focus on some open neighbourhood (portion) of three dimensional space and suppose that a number of physical pavings of the kind just discussed have been constructed in this region. Let (α, β, γ) and $(\bar{\alpha}, \bar{\beta}, \bar{\gamma})$ denote the physical coordinates associated with two of these grids. One can then *empirically* determine the coordinate transformation

$$\begin{aligned}
\bar{\alpha} &= \bar{A}(\alpha, \beta, \gamma)\\
(2.10)\quad \bar{\beta} &= \bar{B}(\alpha, \beta, \gamma)\\
\bar{\gamma} &= \bar{C}(\alpha, \beta, \gamma)
\end{aligned}$$

and its inverse. Newton–Galilei theory informs us that

PROPOSITION 2.11. *Euclideanness.* There exists exactly one one-parameter[4] family of equivalence classes such that, for any two physical coordinate charts (α, β, γ) and $(\bar{\alpha}, \bar{\beta}, \bar{\gamma})$ constructed in the manner presented above (with a given PCHILC-class chosen as the standard unit), the coordinate transformation (A, B, C) and its inverse are similar to Euclidean transformations; that is, the transformation (A, B, C) has the form

$$(2.11)\quad \begin{bmatrix} \bar{\alpha} \\ \bar{\beta} \\ \bar{\gamma} \end{bmatrix} = SR(\psi, \theta, \phi)S^{-1} \begin{bmatrix} \alpha \\ \beta \\ \gamma \end{bmatrix} + S \begin{bmatrix} a \\ b \\ c \end{bmatrix},$$

where $R\,(\psi, \theta, \phi)$ is a 3-dimensional rotation matrix, (a, b, c) determines the translation that relates the origins of the two systems and S is a nondegenerate, upper triangular matrix (6 parameters) that is the same for every such transformation.

The empirical procedure used to test the validity of the proposition is the following. Many physical coordinate systems based on a particular PCHILC-class of rods are constructed in a given region of space. For each pair of charts, the transformation relating them is measured. It will *turn out* that this transformation is linear; that is, it has the form

$$(2.12)\quad \begin{bmatrix} \bar{\alpha} \\ \bar{\beta} \\ \bar{\gamma} \end{bmatrix} = Z \begin{bmatrix} \alpha \\ \beta \\ \gamma \end{bmatrix} + \begin{bmatrix} \zeta \\ \eta \\ \xi \end{bmatrix},$$

where the elements of the matrix Z and ζ, η and ξ are empirically determined. One obtains in this way a collection $\{Z\}$ of such empirically determined matrices. It is claimed that there exists a single upper triangular, invertible matrix S such that each element of the transformed collection $\{S^{-1}ZS\}$ is a rotation; that is, for each element, there is a rotation R such that $S^{-1}ZS = R$. Since rotations satisfy the constraint $R^T R = I$, each of the elements yields a constraint on the six components of S, namely,

$$(2.13) \quad S^T Z^T S^{-1T} S^{-1} ZS = I,$$

where the elements of Z are empirically known. One can use various analytical techniques, such as the method of least squares, to determine the best values for the elements of S and the degree of confidence or goodness of fit of the hypothesis to the data. If this coordinate independent analysis of the data succeeds, then one can conclude that there exists an Euclidean Lie pseudogroup structure or an $\mathbb{E}(\mathbb{R}^3)$-structure on the given portion of space; moreover, the coordinates defined by

$$(2.14) \quad \begin{bmatrix} x \\ y \\ z \end{bmatrix} = S^{-1} \begin{bmatrix} \alpha \\ \beta \\ \gamma \end{bmatrix} = \begin{bmatrix} \alpha + \gamma/2 \\ \beta + \gamma/2 \\ \sqrt{2}/2\,\gamma \end{bmatrix}$$

are adapted to the flat 3-dimensional Euclidean metric which is therefore given by

$$(2.15) \quad ds^2 = dx^2 + dy^2 + dz^2$$

with respect to the physical coordinates (x, y, z). The mathematical justification for this last claim is presented in the next section.

2.3. Euclidean Pseudogroup Structure on n-Dimensional Space

There are several methods used by mathematicians to characterize structures on manifolds. The method that employs Lie pseudogroups is the most suitable in the context of the analysis of flat Euclidean structure presented above. The following definition is an appropriately restricted version of a definition that appears in Kobayashi and Nomizu [22].

DEFINITION 2.12. *Lie Pseudogroup of Transformations on* $\mathbb{R}^n$. A Lie pseudogroup of transformations on $\mathbb{R}^n$ is a set Γ of local diffeomorphisms such that

(1)   Each $f \in \Gamma$ is a local diffeomorphism $f: A \to B$ from an open set $A = $ domain $f \subseteq \mathbb{R}^n$ to an open set $B = f_+(A) = $ range f.

(2)   If $f \in \Gamma$ and C is an arbitrary open subset of $A = $ domain f, then the restriction $f|_C \in \Gamma$.

(3)   Let $A = \cup_i A_i$ where each A_i is an open subset of $\mathbb{R}^n$. A local diffeomorphism $f: A \to B$, where $B \subseteq \mathbb{R}^n$, is an element of Γ provided that for each i, $f|_{A_i} \in \Gamma$.

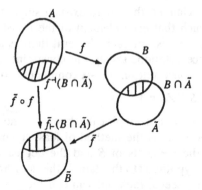

Figure 3. Lie pseudogroup condition 6.

(4)   For every open set A of $\mathbb{R}^n$, the identity map $I: A \to A$ is an element of Γ.

(5)   If $f \in \Gamma$, then $f^{-1} \in \Gamma$.

(6)   If $f \in \Gamma$, $\tilde{f} \in \Gamma$, $f: A \to B$ and $\tilde{f}: \tilde{A} \to \tilde{B}$ and if $B \cap \tilde{A} \neq \emptyset$, then $\tilde{f} \circ f: f^{-1}(B \cap \tilde{A}) \to \tilde{f}_+(B \cap \tilde{A})$ is an element of Γ. This condition is illustrated in Figure 3.

If one restricts to the appropriate subsets, condition (6) of Definition 2.12 simply asserts that if $f_1 \in \Gamma$ and $f_2 \in \Gamma$ and range $f_1 =$ domain f_2, then $f_2 \circ f_1 \in \Gamma$.

Euclidean geometry is characterized by the Euclidean Lie Pseudogroup on $\mathbb{R}^n$ denoted by $\mathbb{E}(\mathbb{R}^n)$. Let $A \subseteq \mathbb{R}^n$ be any open neighbourhood of $\mathbb{R}^n$, and let f be any Euclidean transformation on $\mathbb{R}^n$. Set $B = f_+(A)$. Then, $f|_A: A \to B$ is an element of $\mathbb{E}(\mathbb{R}^n)$; moreover, every element of $\mathbb{E}(\mathbb{R}^n)$ is obtained in this way. It is easy to check that $\mathbb{E}(\mathbb{R}^n)$ is a Lie pseudogroup.

Let M be a differentiable manifold and let Γ be a Lie pseudogroup on $\mathbb{R}^n$. Then, a Γ-structure on M is determined by an atlas for M that is compatible with Γ. Again, the following definition is a specialization of that found in Kobayashi and Nomizu [22].

DEFINITION 2.13. An atlas for a differentiable manifold M that is compatible with a Lie pseudogroup Γ of transformations is an atlas $\{(U_i, x_i)\}$ such that whenever $U_i \cap U_j \neq \emptyset$, the mapping

(2.16) $x_j \circ x_i^{-1}: x_{i+}(U_i \cap U_j) \to x_{j+}(U_i \cap U_j)$

is an element of Γ. A complete atlas of M compatible with Γ is an atlas of M compatible with Γ that is not contained in any other atlas of M. A Γ-structure on M is a complete atlas compatible with Γ.

REMARK 2.14. Let $\mathscr{A} = \{(U_i, x_i)\}$ be an atlas of M compatible with Γ. Let $\tilde{\mathscr{A}}$ be the atlas consisting of all charts (U, x) such that

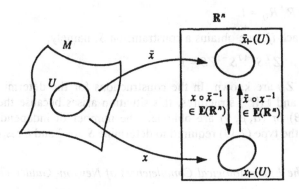

Figure 4. Euclidean Lie pseudogroup structure.

$$x \circ x_i^{-1} : x_{i\vdash}(U_i \cap U) \to x_{\vdash}(U_i \cap U)$$

is an element of Γ. Then, $\tilde{\mathscr{A}}$ is a complete atlas compatible with Γ.

We now specialize the above definition of a Γ-structure on M to the case of a Euclidean pseudogroup or $\mathbb{E}(\mathbb{R}^n)$-structure on a portion (open neighbourhood) of M, that is, a flat Euclidean metric on a portion of M. In the Appendix Section A.2, we show how such a structure determines a cross section of the associated fiber bundle $O(\mathbb{R}^n) \backslash \mathscr{L}^*(M)$ of $O(\mathbb{R}^n)$-related coframes, that is, a (flat) Riemannian structure on M. Since only a portion U of M is considered, all of the charts in the atlas $\mathscr{A}$ for U that are compatible with $\mathbb{E}(\mathbb{R}^n)$ may be regarded as defined on U. For any two charts (U, x) and $(U, \tilde{x})$ in $\mathscr{A}$, the maps $\tilde{x} \circ x^{-1}$ and $x \circ \tilde{x}^{-1}$ belong to $\mathbb{E}(\mathbb{R}^n)$. The situation is illustrated in Figure 4.

REMARK 2.15. Although an $\mathbb{E}(\mathbb{R}^n)$-structure on U is defined by an atlas $\mathscr{A} = \{(U, x_i)\}$ for U such that each element of the family $\mathscr{X} = \{X_{ij} = x_i \circ x_j^{-1}\}$ of transformations belongs to $\mathbb{E}(\mathbb{R}^n)$, such a structure could also be described with respect to a nonadapted coordinated system; that is, an atlas $\mathscr{B} = \{(U, y_i)\}$, where for some invertible linear[5] transformation $S: \mathbb{R}^n \to \mathbb{R}^n$, $y_i = S \circ x_i$ for every i, also determines the same $\mathbb{E}(\mathbb{R}^n)$-structure. The corresponding family of transformations is then $\mathscr{Y} = \{Y_{ij} = S \circ X_{ij} \circ S^{-1}\}$. Denote the matrix of the linear part of the transformation Y_{ij} by Z_{ij}. Since an arbitrary element of $GL(\mathbb{R}^n)$ can be written as the product of an upper triangular matrix and an element of $O(\mathbb{R}^n)$, one can take S to be upper triangular. For each of the matrices Z_{ij}, the matrix R_{ij} of the homogeneous part of X_{ij} is a rotation

(2.17) $R_{ij} = S^{-1} Z_{ij} S.$

Since the R_{ij} are rotations,

51

(2.18) $R_{ij}^T R_{ij} = I$.

Thus for each ij, one obtains a constraint on S, namely,

(2.19) $S^T Z_{ij}^T S^{-1T} S^{-1} Z S = I$,

where the Z_{ij} are known. In the constructions for the determination of the $\mathbb{E}(\mathbb{R}^2)$ and $\mathbb{E}(\mathbb{R}^3)$ structures, this situation arises because the coordinates (α, β) or (α, β, γ) are oblique. The number of independent constraints of the type (2.19) required to determine S are 3 and 6 respectively.

2.4. *The Epistemological Completeness of Newton–Galilei Theory*

In Section 3 of [1, Coleman and Korté, appearing in this issue], we presented a number of general principles and/or desiderata that any reasonable epistemology of geometry should conform with. The constructions and analyses presented above for the empirical determination of Euclidean geometry adhere to these principles.

PROPOSITION 2.16. *Epistemic Completeness of Newton–Galilei Theory.* Newton–Galilei theory is epistemologically complete with respect to its spatial metric.

Proof: Newton–Galilei theory informs us that the appropriate measuring device for the determination of spatial distance is a rigid rod and that rigid rods must satisfy at least the Conditions 2.3 and 2.4 for a HILC-rod. Whether or not a rod satisfies the theoretical criteria to be a HILC-rod can, however, be tested in a way that is independent of all the post-differential-topological aspects of the theory because the test requires only the observation of sufficient number of local spatial coincidences.

The theory also says that space is Euclidean. It is a theorem that space is Euclidean iff there exist a *regular* paving, either an equilateral paving if space is two-dimensional or a regular octahedral-tetrahedral paving if space is three-dimensional. In addition the theory tells us that the coordinates defined by the *regular* paving in the manner described above are adapted to the Euclidean geometry. Although it is not known at this stage whether or not HILC-rods are rigid, we nevertheless proceed with the construction of a physical paving (either triangular or octahedral-tetrahedral). The test as to whether or not a (possibly not regular) physical paving of the relevant type is obtained depends only on the local differential topology, that is, on whether or not the rods fit together at the appropriate vertices. If the rods do fit together, we shall say that the property of *paving-closer* is satisfied and call the associated rods PCHILC-rods. If paving-closure is satisfied, the paving provides a physical coordinate system. Because we have made no assumption regarding congruence of the rods under transport, aside from the HILC-conditions which are testable on the basis of the local differential topology, we cannot, however, conclude that the physical coordinate system so constructed is adapted to a Euclidean geometry; that is, we cannot at this stage invoke the proposition of Euclideanness 2.10 or 2.11.

At this point, one can proceed to construct a number of these coordinate systems on U and to measure the coordinate transformations that relate them by determining for a number of points with coordinates $\{(\bar{\alpha}_I, \bar{\beta}_I, \bar{\gamma}_I)\}$, $I \in \{1, 2, \ldots, N\}$ the corresponding coordinates $\{(\alpha_I, \beta_I, \gamma_I)\}$, $I \in \{1, 2, \ldots, N\}$. One can then determine mathematically whether or not these empirically determined coordinate transformations are linear by various analytic techniques such as the methods of least squares and goodness of fit. If the postulates of Newton–Galilei theory are correct, then there will exist classes of PCHILC-rods for which the transformations do turn out to be linear. Then one obtains a large number of matrices $\{Z\}$ that describe the homogeneous part of these linear transformations. One can then test the hypothesis that there exists a unique upper triangular matrix S such that each member of the set of matrices $\{S^{-1}ZS\}$ is a rotation matrix again by analytic methods such as the method of least squares and goodness of fit as discussed above. If the postulates of Newton–Galilei theory hold, this analysis will be successful and the Euclidean metrics of space are given by (2.9) and (2.15) respectively. Although some specific set of physical coordinate systems and the empirically determined coordinate transformations between them are used in the above analyses the spatial metric so obtained is coordinate independent. ■

3. A NONCONVENTIONAL DISCOVERY PROCEDURE FOR THE SPACETIME METRIC IN EINSTEIN–MAXWELL THEORY

3.1. *Theoretical Completeness of Einstein–Maxwell Theory*

The ontological domain of Einstein–Maxwell theory contains among other things the spacetime metric field, the electromagnetic field and various kinds of particles. In early work, Kretschmann [21] and Weyl [25] showed how a measurement procedure for the spacetime metric could be based on the motions of neutral massive monopoles and electromagnetic waves. In particular, Weyl proved a theorem which states that the conformal structure and the projective structure of a space equipped with a (pseudo) Riemannian structure together determine that structure. He also remarked that it would therefore be possible to determine that spacetime metric provided that the motions of neutral massive monopoles and electromagnetic fields could be followed, thereby essentially[6] establishing the *theoretic completeness* of Einstein–Maxwell theory with respect to its geometric fields. Despite their work, epistemic analyses of the measurement of the spacetime metric were for many years still based on the use of rigid rods and ideal clocks (entities that are not part of the ontological domain of

Einstein–Maxwell theory) an approach that does not lead to a measurement procedure that makes use only of criteria, procedures and analyses that are strictly differential-topological in character. In this section, we show in detail how the work of of Kretchmann and Weyl can be extended to show that Einstein–Maxwell theory is *epistemically complete* with respect to its geometric fields. We shall emphasize the conceptual details rather than the technical details which have been presented elsewhere [2–14].

In Section 3.3 of [1, Coleman and Korté, appearing in this issue] we give a brief historical overview of the development of results pertaining to the theoretic completeness of Einstein–Maxwell theory with respect to its geometric structure. The discussion in that section and the preceding remarks can be summarized by the following proposition.

PROPOSITION 3.17. *Theoretical Completeness of Einstein–Maxwell Theory*. Einstein–Maxwell theory is theoretically complete with respect to its spacetime metric.

Proof:

(1)   The ontological domain of Einstein-Maxwell theory includes among other things the space-time metric field, the electromagnetic field and various kinds of particles.

(2)   The theory provides within its ontological domain a theoretical account of physical coordinate charts that are compatible with the local differential topology that underlies the theory; for example, see the account presented in Section 2.3 of [1, Coleman and Korté, appearing in this issue] and illustrated in Figure 3 of that paper. We have also described elsewhere [13] a physical coordinate chart, a radar-station chart, modeled on modern radar installations.

(3)   The theory states
   (a) that the motion of the electromagnetic waves of any frequency is governed by the causal-conformal structure associated with the spacetime metric, and
   (b) that if these motions could be followed in sufficient detail, they would reveal the causal-conformal structure.

(4)   The theory states that
   (a) the motion of neutral massive monopoles is governed by the projective structure associated with the spacetime metric, and
   (b) that if these motions could be followed in sufficient detail, they would reveal the projective structure.

(5)   **Weyl's Theorem:** The causal-conformal and projective structures determined by a spacetime metric together determine that spacetime metric.   ■

54

REMARK 3.18. While it follows from items (3b) and (4b) that the ability to follow the motions of electromagnetic waves and neutral massive monopoles suffices for the determination of the conformal and projective structures, without the theoretical result stated in item (5), it does not follow that these motions suffice for the determination of the spacetime metric.

3.2. *Epistemological Completeness of Einstein–Maxwell Theory with Respect to its Conformal and Projective Structure*

As is indicated in the list of items of Proposition 3.17 concerning the theoretic completeness of Einstein–Maxwell theory with respect to its post-differential-topological, geometric structures, it is necessary to show that Einstein–Maxwell theory is epistemically complete with respect to both its conformal structure and its projective structure.

3.2.1. *Epistemological Completeness with Respect to the Conformal Structure*

We have discussed at length elsewhere [13] how the propagation of electromagnetic waves with respect to a physical coordinate system can be determined using a sufficient number of probes that can receive and emit tagged electromagnetic waves that are used for the purpose of assigning coordinates to the emission and reception events. Here, we simply point out that this tracking procedure can be used to check at the differential-topological level a number of relevant claims about the propagation of electromagnetic waves, particularly the fact that the propagation in vacuo of electromagnetic waves of *all frequencies* is governed by one and the same causal conformal structure; moreover, at each event q corresponding to physical coordinates $x^i_\mathrm{p}(q)$, a 'curve fitting' analysis of the data yields the physical descriptors $g_{ij}(x^i_\mathrm{p}(q))$ with respect to the physical coordinates x^i_p of the coordinate independent conformal structure

$$(3.20) \quad g_{ij}(x^i_\mathrm{p}(q))d_q x^i_\mathrm{p} \otimes d_q x^j_\mathrm{p}$$

at q. The descriptors $g_{ij}(x^i_\mathrm{p})$ are of course measured at many different events throughout a region U of spacetime typically along a number of worldline paths through the region U. From this data, the so called conformal connection coefficients $K^i_{jk}(x^i_\mathrm{p})$ can be computed mathematically by interpolation and differentiation techniques.

In the absence of disturbances (*in vacuo*), all electromagnetic waves of whatever source are governed by the same conformal structure, a fact that can be tested in the manner just indicated; consequently, complicated sorting and testing procedures of the kind discussed below in connection with the measurement of the projective structure are not required for the measurement of the conformal structure.

3.2.2. *Epistemological Completeness with Respect to the Projective Structure*

To establish the epistemic completeness of Einstein–Maxwell theory with respect to its projective structure requires a more complicated set of criteria and procedures because the theory informs us that *even in the absence of disturbances*, there are different kinds of massive particles with different behaviours such as nonmonopoles, charged monopoles and neutral monopoles. The theory informs us that the motion of a massive particle that belongs to a given monopole class is governed by an equation-of-motion structure Ξ called a directing field [2, 5–8]. A portion of the worldline path of such a particle is described by an equation

$$(3.21) \quad x^\alpha = \xi^\alpha(t)$$

where $\xi^\alpha(t)$ satisfies a differential equation of the form

$$(3.22) \quad D^2\xi^\alpha(t) = \Xi^\alpha(t, \xi^\alpha(t), D\xi^\alpha(t)),$$

where $D \overset{\text{def}}{=} d/dt$. A solution of such a differential equation is uniquely determined by the values $(\xi^\alpha(t_0), D\xi^\alpha(t_0))$ at any time $t = t_0$. For a non-monopole, on the other hand, the values $(\xi^\alpha(t_0), D\xi^\alpha(t_0))$ do not in general suffice to determine the future path of the particle because this path depends in general on additional quantities that specify its initial orientation. These theoretical considerations lead to the following differential-topological criterion for monopoles of a given class.

CRITERION 3.19. *Monopole Criterion.* If two monopoles of the same class are launched at nearly the same event with nearly the same 3-velocity, then under all external physical conditions, their (future) worldline paths will be nearly the same.

Intuitively, this criterion means that any two monopoles of the *same* class can either fly in formation or play tag in an *arbitrary* external force field. With respect to a physical coordinate system x^i_p, one would choose for the two particles initial data of the form

$$(3.23) \quad (t_0, \xi^\alpha_0, \xi^\alpha_1) \quad \text{and} \quad (t_0, \xi^\alpha_0 + \delta\xi^\alpha_0, \xi^\alpha_1)$$

for the flying-in-formation case and initial data of the form

$$(3.24) \quad (t_0, \xi^\alpha_0, \xi^\alpha_1) \quad \text{and} \quad (t_0 + \delta t_0, \xi^\alpha_0, \xi^\alpha_1)$$

for the playing-tag case, where $\xi^\alpha(t)$ describes the worldline path of one of the particles, $\xi^\alpha_0 = \xi^\alpha(t_0)$, $\xi^\alpha_1 = D\xi^\alpha(t_0)$, and both δt_0 and $\delta\xi^\alpha_0$ are 'small' in the differential-topological sense of a limiting sequence of experiments. This criterion is based on the theorems concerning the stability of the solutions of the differential equation (3.22) under 'small' variations of the initial data.

It is important to emphasize that the sorting must be rigorous; that is, the particles should be tested against each other for a large sample of different initial conditions and in as many different external force fields as can be found or generated. The reason for this requirement is that degeneracies can occur. Consider two different monopoles, for example, a neutral monopole with mass m and a monopole with mass m and electromagnetic charge q. If in some region the electromagnetic field is zero, these different monopoles can fly in formation or play tag regardless of the initial conditions, but they will not do so if the electromagnetic field is nonzero. Similarly, a multipole moment is typically revealed only in the presence of some *inhomogeneous* force field; for example, a particle with a magnetic dipole moment couples to the gradient of the magnetic field and may behave like a monopole in the absence of an inhomogeneous magnetic field. In summary, if there are a number of monopole classes, then the members of a given class are characterized by the fact that any two members of that class can be made to fly in formation or play tag for arbitrary initial conditions under the influence of all external force fields that can be either found or generated.

If one has a sufficient number of monopoles of a given monopole class, the differential-topological procedure for determining the directing field Ξ that governs their motions is the following.[7]

(1)   A large number of these monopoles are launched with a wide variety of initial conditions in such a way that their worldline paths traverse a region U of spacetime.

(2)   Each of these monopoles is tracked with respect to a physical system of coordinates $x^i_{\underset{P}{}}$ and the equations

$$(3.25) \quad x^\alpha_{\underset{P}{}} = \xi^\alpha_{\underset{P}{A}}(t_{\underset{P}{}}), \quad A \in \{1, \ldots, N\}$$

of their world line paths are empirically determined using appropriate 'curve-fitting' techniques.

(3)   By differentiating the Equations (3.25) twice, one obtains the lifts of these paths, namely,

$$x^\alpha_{\underset{P}{}} = \xi^\alpha_{\underset{P}{A}}(t_{\underset{P}{}})$$

$$(3.26) \quad \xi^\alpha_{\underset{P}{1}} = D\xi^\alpha_{\underset{P}{A}}(t_{\underset{P}{}})$$

$$\xi^\alpha_{\underset{P}{2}} = D^2\xi^\alpha_{\underset{P}{A}}(t_{\underset{P}{}}), \quad A \in \{1, \ldots, N\}.$$

Each of these paths lies on a 7-dimensional submanifold

of a 10-dimensional manifold locally coordinatized by $(t, x^\alpha, \xi_1^\alpha, \xi_2^\alpha)$, where ξ_1^α and ξ_2^α are the coordinates corresponding to 3-velocity and 3-acceleration respectively.

(4)   The equation for the 7-dimensional submanifold on which each
of the paths (3.26) lies is

$$(3.27) \quad \xi_2^\alpha = \Xi^\alpha(t, x^\alpha, \xi_1^\alpha).$$

The functions Ξ^α can be determined by means of suitable
interpolation techniques. It should be noted that although the
functions Ξ^α that describe Ξ refer to a particular coordinate
system, the directing field Ξ itself is a coordinate independent
geometric-object field.

(5)   A directing field is geodesic if and only if the functions that
describe it are cubic in the 3-velocity coordinates with leading
term proportional to the 3-velocity in some (and hence in every)
coordinate system. This criterion is clearly coordinate independent since it does not matter in which coordinate system it is
applied. Hence, to empirically determine whether or not a
measured directing field described by functions Ξ^α with respect
to a system of physical coordinates x^i is geodesic, one need
only differentiate the measured functions $\Xi^\alpha(t, x^\alpha, \xi_1^\alpha)$ with
respect to the variables ξ_1^α and show that it is a cubic with
respect to the variables ξ_1^α with the purely cubic term proportional to ξ_1^α. The criterion for geodesicity is, therefore,
testable at the level of the local differential topology.

CRITERION 3.20. *Cubic Criterion.* A directing field Ξ is geodesic and
is denoted by Π if and only if

$$(3.28) \quad \Pi^\alpha(x^i, \xi_1^\alpha) = \xi_1^\alpha(\Pi_{\rho\sigma}^0(x^i)\xi_1^\rho\xi_1^\sigma + 2\Pi_{0\rho}^0(x^i)\xi_1^\rho + \Pi_{00}^0(x^i))$$

$$-(\Pi_{\rho\sigma}^\alpha(x^i)\xi_1^\rho\xi_1^\sigma + 2\Pi_{0\rho}^\alpha(x^i)\xi_1^\rho + \Pi_{00}^\alpha(x^i)),$$

where the projective coefficients $\Pi_{jk}^i(x^i)$ satisfy $\Pi_{jk}^i(x^i) = \Pi_{kj}^i(x^i)$ and
$\Pi_{jk}^j(x^i) = 0$. The projective coefficients $\Pi_{jk}^i(x^i)$ uniquely determine and
are uniquely determined by a projective structure relative to a choice of
coordinates x^i. Clearly, from the empirically determinable functions
$\Pi^\alpha(x^i, \xi_1^\alpha)$ of a geodesic directing field, one can uniquely determine the
physical descriptors $\Pi_{jk}^i(x^i)$ of the projective structure in the region U of
spacetime.

    There remain two simple, coordinate independent, mathematical criteria which, if satisfied, guarantee that the physical descriptors $\underset{p}{g}_{ij}(\underset{p}{x}^i)$, $\underset{p}{K}^i_{jk}(\underset{p}{x}^i)$ and $\underset{p}{\Pi}^i_{jk}(\underset{p}{x}^i)$ determine the physical descriptors $\underset{p}{g}_{ij}(\underset{p}{x}^i)$ of the spacetime metric.

CRITERION 3.21. *EPS-Compatibility Criterion.* The conformal and projective structures together determine a Weyl structure described locally by physical descriptors $\underset{p}{g}_{ij}(\underset{p}{x}^i)$ and $\underset{p}{\Gamma}^i_{jk}(\underset{p}{x}^i)$.

The algebraic condition for this criterion is due to Ehlers, Pirani and Schild [18].

CRITERION 3.22. *Streckenkrümmung Criterion.* The tensorial Streckenkrümmung condition

$$(3.29) \quad \underset{p}{\Gamma}_{j,k}(\underset{p}{x}^i) - \underset{p}{\Gamma}_{k,j}(\underset{p}{x}^i) = 0,$$

where $\underset{p}{\Gamma}_k(\underset{p}{x}^i) = \underset{p}{\Gamma}^j_{jk}(\underset{p}{x}^i)$, ensures that the Weyl structure reduces to a pseudo-Riemannian structure locally described by physical descriptors $\underset{p}{g}_{ij}(\underset{p}{x}^i)$.

REMARK 3.23. It could, in principle, turn out that more than one of the monopole classes was geodesic. If this were so, however, it would mean that the Einstein–Maxwell theory was wrong in some important respect; moreover, this fact could be discovered by means of the differential-topological criteria and procedures discussed above.

REMARK 3.24. If the coordinate independent differential-topological criterion for EPS-compatibility of the conformal and projective structures were not satisfied, then the massive monopoles governed by the projective structure could break the light barrier!

REMARK 3.25. If the coordinate independent, differential-topological Streckenkrümmung condition were not satisfied, then the Einstein–Maxwell theory would again be wrong in some important respect and such phenomena as second clock effects would exist.

3.3. *Epistemological Completeness of Einstein–Maxwell Theory with Respect to its Metric Structure*

The preceding discussion is summarized by the following proposition.

PROPOSITION 3.26. *Epistemological Completeness of Einstein–Maxwell*

Theory. Einstein–Maxwell theory is epistemically complete with respect to its spacetime metric.

Proof: First, Einstein–Maxwell theory is theoretically complete for the reasons stated in the proof of Proposition 3.17. According to item (5) of the proof of Proposition 3.17, Einstein–Maxwell theory is epistemically complete with respect to its spacetime metric if and only if it is epistemically complete with respect to the conformal and projective structures determined by that metric.

SUBPROPOSITION 3.27. Einstein–Maxwell theory is epistemically complete with respect to its conformal structure.
   Proof:

   (1)   The theory informs us that the propagation of all electromagnetic waves are governed by the same conformal structure and that these motions suffice to reveal the conformal structure provided they can be followed in sufficient detail.
   (2)   There exist differential-topological procedures for the empirical determination of the propagation of electromagnetic waves with respect to physical coordinate systems.
   (3)   By tracking electromagnetic waves emitted from a point q, one can determine from a coordinate independent analysis of the data that the propagation of all electromagnetic waves is governed by the same causal-conformal structure and one can determine the physical descriptors $g_{ij}(x^i(q))$ of that structure at q. By repeating the process at a large number of points in a region U of spacetime, one can determine the coordinate independent conformal structure throughout U. ▼

The theory informs us that particles can be either monopoles or non-monopoles and that monopoles may be of various kinds; moreover, mathematical analysis informs us that the motions of each kind of monopole are governed by a directing field and that if these motions could be followed in sufficient detail, they would reveal that directing field.

SUBPROPOSITION 3.28. Einstein–Maxwell theory is epistemically complete with respect to its (geometric-object) directing-field structures.
   Proof: First, there exists a differential-topological Criterion 3.19 on the basis of which particles can be sorted into nonmonopoles and monopoles of different kinds. Second, given a sufficient number of monopoles of a particular class, there exist differential-topological procedures for determining the physical descriptors $\Xi^\alpha(t, x^i, \xi_1^\alpha)$ of the directing field structure. ▼

The theory informs us that of all the directing fields, one and only one is a geodesic directing field, that is, a projective structure.

SUBPROPOSITION 3.29. Einstein–Maxwell theory is epistemically complete with respect to its unique projective structure.

Proof: One need only apply the differential-topological and coordinate independent cubicness Criterion 3.20 to the physical descriptors $\underset{p}{\Xi}{}^{\alpha}$ of each of the directing fields to determine whether or not one and only one of them passes the test. If one and only one of them does pass the test, then the physical descriptors $\underset{p}{\Pi}{}^{i}_{jk}(\underset{p}{x}{}^{i})$ of the unique projective structure may be obtained be differentiation of the physical descriptors of the geodesic directing field with respect to the variables $\underset{p}{\xi}{}^{\alpha}_{1}$. ▼

Finally, to show that the conformal and projective structures determine a (unique up to a constant positive factor) pseudo-Riemannian metric one need only check that the differential-topological and coordinate independent conditions for EPS-compatibility and for the absence of second clock effects (the vanishing of the Streckenkrümmung tensor) are satisfied. ■

APPENDIX A. GEOMETRIC FIELDS AND THE EQUIVALENCE OF THE LIE-PSEUDO GROUP AND G-STRUCTURE DESCRIPTIONS OF EUCLIDEAN GEOMETRY

The purpose of this appendix is two-fold. First, we provide a characterization of the subset of geometric-object fields called geometric fields. Second, we show that the Lie-pseudogroup description of Euclidean geometry used in this paper is equivalent to the more familiar (flat) G-structure or vierbein description of Euclidean geometry. The presentation given here focuses on the special case that is relevant to the concerns of this paper. The proofs that appear in the mathematical literature [19, 20, 24] are applicable to the most general case and are consequently not readily accessible to most readers.

A.1. *Geometric Fields*

Both geometric fields and geometric-object fields are defined over or exist with respect to a fundamental base space which we denote by M. The space M is an n-dimensional differentiable manifold usually called spacetime or space-and-time depending on the theory in question. Usually, $n = 4$ although higher dimensional worlds are sometimes considered. There is a great variety of geometric objects. Vectors and co-vectors are familiar examples. A geometric object Ω is attached to a point $p \in M$ and is described by a finite system of numbers Ω_I with respect to a local coordinate chart $(U, x)_p$ for a neighbourhood U of p. Under a change of coordi-

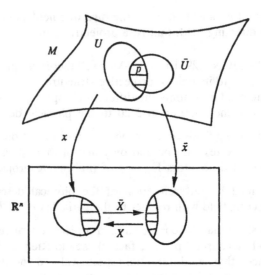

Figure 5. Coordinate transformation.

nate chart from $(U, x)_p$ to $(\bar{U}, \bar{x})_p$, the geometric object is re-described by a new system of numbers $\bar{\Omega}_I$ that are determined by the numbers Ω_I and the values of a finite number of derivatives of the transformation functions $\bar{X} = \bar{x} \circ x^{-1}: x_+(\bar{U} \cap U) \rightarrow \bar{x}_+(\bar{U} \cap U)$ at $x(p)$. The order of the highest partial derivative of $\bar{X}$ that appears in the expressions for the $\bar{\Omega}_I$ is called the *order* of the geometric object Ω.

All of the geometric objects Ω at a point $p \in M$ of a given kind together form a differentiable manifold $F(M_p)$ the dimension l of which is the number of components Ω_I that describe the geometric object with respect to the chart $(U, x)_p$. The (disjoint) union $F(M) = \sqcup_p F(M_p)$ is a differentiable manifold of dimension $n + l$. The surjective submersion $\pi : F(M) \rightarrow M$ is smooth. The structure

(A.30) $\mathcal{F}(M) = \langle F(M), \pi, M, F(\mathbb{R}^n_0) \rangle.$

is called a fiber bundle over the base space M. The manifold $F(M)$ is called the total space, π is called the projection and the manifold $F(M_p)$ denotes the fiber over $p \in M$.

The typical fiber $F(\mathbb{R}^n_0)$ is the set of geometric objects at the standard point $\mathbf{0}$ of the standard coordinate space $\mathbb{R}^n$ that are roughly speaking defined in the 'same' way as the geometric objects in $F(M_p)$; for example, if the geometric objects are tangent vectors, the typical fiber is the n-dimensional manifold of vectors tangent to $\mathbb{R}^n$ at $\mathbf{0}$. A *cross section* of the bundle $\mathcal{F}(M)$ is a smooth map $\sigma : M \rightarrow F(M)$ such that $\pi \circ \sigma = \text{Id}_M$. It can happen that there are two fiber bundles $\mathcal{F}^A(M)$ and $\mathcal{F}^B(M)$ such that there is a surjective submersion

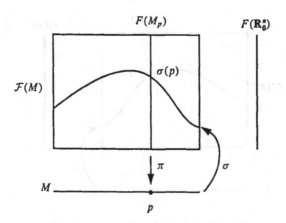

Figure 6. The associated fiber bundle $\mathcal{F}(M)$.

(A.31) $\pi_B^A: F^B(M) \to F^A(M)$;

so that $F^B(M)$ is the total space of a fiber bundle over $F^A(M)$. In this case, one also calls a smooth map $\sigma: F^A(M) \to F^B(M)$, such that $\pi_B^A \circ \sigma = \mathrm{Id}_{F^A(M)}$, a cross section. We shall call both types of cross sections geometric-object fields. Among those geometric-object fields determined by the second type of cross section are the equation-of-motion structures for massive monopoles. The term 'massive monopole' is used in the sense of 'unstructured test bodies'.

A geometric field is a special type of geometric-object field called a G-structure on M. Most geometric fields are first order G-structures which are reductions of the principal bundle of (first order) coframes. A coframe is defined as follows. Denote by $\mathscr{C}$ the set of charts $(V, y)_p$ where V is a neighbourhood of p. Let $(U, x)_p$ be any element of $\mathscr{C}$. Define an equivalence relation on $\mathscr{C}$ by the requirement that $(V_\alpha, y_\alpha)_p \sim (V_\beta, y_\beta)_p$ iff $(y_\alpha - y_\alpha(p)) \circ x^{-1}|_{x \vdash (V_\alpha \cap U)}$ and $(y_\beta - y_\beta(p)) \circ x^{-1}|_{x \vdash (V_\beta \cap U)}$ have the same first partial derivatives at $x(p)$. That the equivalence relation does not depend on the choice of the chart $(U, x)_p$ is a consequence of the chain rule. An equivalence class $[(V, y)_p]$ is called a linear coframe at $p \in M$. The partial derivatives

(A.32) $H_j^i = \dfrac{\partial (y^i - y^i(p)) \circ x^{-1}}{\partial x^j}(x(p))$

are the components of the coframe with respect to the chart $(U, x)_p$. Under a change of chart from $(U, x)_p$ to $(\bar{U}, \bar{x})_p$, the components transform according to

(A.33) $\bar{H}_j^i = H_r^i X_j^r$,

where the X_j^i are the *first* partial derivatives of the transformation func-

63

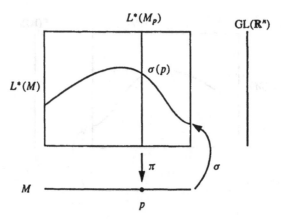

Figure 7. The principal fiber bundle $\mathcal{L}^*(M)$.

tions $X^i = x^i \circ \bar{x}^{-1}|_{\bar{x}_{\vdash}(U \cap \mathcal{O})}$ at $\bar{x}(p)$; consequently, the geometric object $[(V, y)_p]$ is of *first* order. Since $(y - y(p)) \circ x^{-1}$ is invertible, the H^i_j are the elements of a non-degenerate $n \times n$ matrix. The set of all coframes at $p \in M$ forms an n^2-dimensional manifold $L^*(M_p)$. The disjoint union $L^*(M) = \bigsqcup_p L^*(M_p)$ is a differentiable manifold of dimension $n + n^2$. The projection $\pi_{L^*}: L^*(M) \to M$ is defined by

(A.34) $\pi([(V, y)_p]) = p$.

Each chart (U, x) for M determines a chart $(L^*(U), (x, H))$ for the portion $L^*(U) = \pi^{-1}(U)$ of $L^*(M)$ over U, where $x([(V, y)_p]) = (x^i)$ and $H([(V, y)_p]) = H^i_j$. The group $GL(\mathbb{R}^n)$ of invertible $n \times n$ matrices acts on $L^*(M)$ in a way that preserves the submanifolds $L^*(M_p)$ each of which is diffeomorphic to $GL(\mathbb{R}^n)$ which is the typical fiber of the principal fiber bundle of linear coframes

(A.35) $\mathcal{L}^*(M) = \langle L^*(M), \pi_{L^*}, M, GL(\mathbb{R}^n) \rangle$.

It is characteristic of a principal fiber bundle that its typical fiber is a Lie group.

A cross section $\sigma: M \to L^*(M)$ of $\mathcal{L}^*(M)$ determines a very rigid geometric structure because it designates a single linear coframe at each point $p \in M$. For each point $p \in M$, there is, so to speak, one and only one correct first order view or image of the infinitesimal neighbourhood of $p \in M$. Less rigid geometric structures are determined by specifying in a smooth way a class of linear coframes at each $p \in M$. Let G be a closed Lie subgroup of $GL(\mathbb{R}^n)$ such as $SL(\mathbb{R}^n)$, $O(\mathbb{R}^n)$ and $O(\mathbb{R}^{1,n-1})$. By restricting the action of $GL(\mathbb{R}^n)$ on $L^*(M_p)$ to the subgroup, one can partition $L(M_p)$ into equivalence classes of linear coframes such that any two elements of an equivalence class are related by an element of G. The space of equivalence classes of G-related coframes at $p \in M$, denoted by

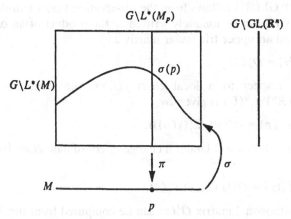

Figure 8. The associated fiber bundle $G\backslash \mathscr{L}^*(M)$.

$G\backslash L^*(M_p)$, is diffeomorphic to the left coset space $G\backslash GL(\mathbb{R}^n)$. The disjoint union $G\backslash L^*(M) = \sqcup_p G\backslash L(M_p)$ is a differentiable manifold which is the total space of the associated fiber bundle of G-related coframes

(A.36) $G\backslash \mathscr{L}^*(M) = \langle G\backslash L^*(M), \pi_{G\backslash L^*}, M, G\backslash GL(\mathbb{R}^n), \mathscr{L}^*(M)\rangle,$

where $G\backslash GL(\mathbb{R}^n)$ is the typical fiber and $\mathscr{L}^*(M)$ is the principal bundle with which $G\backslash \mathscr{L}^*(M)$ is associated. The structure $G\backslash \mathscr{L}^*(M)$ is illustrated schematically in Figure 8.

A cross section $\sigma\colon M \to G\backslash L^*(M)$ defines a geometric field (structure) on M called a G-structure. Consider first the two extreme cases; namely, $G = \{I\}$ and $G = GL(\mathbb{R}^n)$. If $G = \{I\}$, then $G\backslash GL(\mathbb{R}^n) = GL(\mathbb{R}^n)$ and $G\backslash \mathscr{L}^*(M) = \mathscr{L}^*(M)$; consequently, the G-structure is the rigid geometric structure discussed above. The rigidity of this structure is correlated with the 'smallness' of the group (consisting of the identity element alone) that determines the equivalence classes of coframes. In this case, there is *just one* linear coframe in each equivalence class. At the other extreme, the group $G = GL(\mathbb{R}^n)$ is as 'large' as possible, and at each $p \in M$, there is just one equivalence class that contains *all* of the linear coframes at $p \in M$. Such a G-structure is trivial because all first order views (linear coframes) of the infinitesimal neighbourhood of $p \in M$ are regarded as equally correct. There is no constraint imposed at all; consequently, there is no structure.

An intermediate case is that of a Riemannian structure on M corresponding to the group $G = O(\mathbb{R}^n)$. A cross section of $O(\mathbb{R}^n)\backslash \mathscr{L}^*(M)$ determines at each point $p \in M$ an equivalence class of $O(\mathbb{R}^n)$-related linear coframes. Each of the linear coframes in the designated equivalence class $\sigma(p)$ provides a correct first order view of the infinitesimal neighbourhood of $p \in M$ while all of the other linear coframes at $p \in M$ provide views that are more or less distorted. One coordinate system for the coset space $O(\mathbb{R}^n)\backslash GL(\mathbb{R}^n)$ follows from the observation that a non-degenerate $n \times n$ matrix H^i_j can be uniquely written as the product of an orthogonal matrix O^i_j and an upper triangular matrix Σ^i_j

(A.37) $\quad H^i_j = O^i_r \Sigma^r_j.$

Thus, with respect to a local chart (U, x), the local cross section $\sigma_U: U \rightarrow O(\mathbb{R}^n)\backslash L^*(U)$ is given by

(A.38) $\quad \sigma_U(p) = (x^i(p), \Sigma^i_j(x(p))),$

where $\Sigma^i_j(x) = 0$ for $i > j$. Under a change of coordinate chart from (U, x) to $(\bar{U}, \bar{x})$,

(A.39) $\quad \bar{\Sigma}^i_j(\bar{x}) = \bar{O}^i_r(x)\Sigma^r_s(x)X^s_j(\bar{x}),$

where the orthogonal matrix $\bar{O}^i_r(x)$ can be computed from the $X^i_j(\bar{x})$ and the $\Sigma^i_j(x)$. The metric tensor $g_{ij}(x^i)$ determined by the n-bein fields $\Sigma^i_j(x)$ is given by

(A.40) $\quad g_{ij}(x) = \delta_{rs}\Sigma^r_i(x)\Sigma^s_j(x).$

The transformation law (A.39) leads to

(A.41) $\quad \bar{g}_{ij}(\bar{x}) = g_{rs}(x)X^r_i(\bar{x})X^s_j(\bar{x}).$

The $O(\mathbb{R}^n)$-structure formulation of Riemannian geometry has the virtue of emphasizing the role of the microisotropy (microsymmetry) group of the geometry. The description of a Lorentzian structure on M is similar with the group $O(\mathbb{R}^n)$ replaced by the group $O(\mathbb{R}^{1,n-1})$ and with δ_{ij} replaced by the Minkowski metric η_{ij}.

A G-structure is a field that cannot vanish anywhere because at each point a G-structure specifies an equivalence class of coframes and coframes are entities that are non-degenerate by definition. For the case of a flat G-structure, the coframes at different points are correlated in a special way, but they do not vanish. It follows that every point of a manifold M equipped with a G-structure is necessarily occupied.

We shall explain another important feature of a G-structure within the context of a first order G-structure. At any given point $p \in M$, there exist coordinate systems x^i that are adapted to the G-structure at $p \in M$ in the sense that one of the coframes in a $\sigma(p)$ is represented by the identity element of $GL(\mathbb{R}^n)$. These adapted coordinates are related up to first order by an element of G; consequently, the standard representative of the G-structure at $p \in M$ with respect to an adapted coordinate system is the identity element of $GL(\mathbb{R}^n)$.

For the Riemannian case discussed above, $\Sigma^i_j(x(p)) = \delta^i_j$ if x is adapted at $p \in M$. Since the identity element of a group is unique, this notion of adapted coordinate system is nonconventional. There are G-structures of higher order including the second order G-structure associated with a symmetric linear connection and the second order G-structure associated with a projective structure. In addition, lower order G-structures can be prolonged to higher order. The notion of adapted coordinate carries over to these G-structures and is also defined in a similar manner with reference to the identity element of the appropriate higher order structure group.

A.2. The Equivalence of the Lie-Pseudo Group and G-Structure Descriptions of Euclidean Geometry

Let $\{(U, x_i)\}$ be a complete atlas for the $\mathbb{E}(\mathbb{R}^n)$-structure on U as defined in Section 2.3. One can choose *any* one of the charts, say (U, x), in the $\mathbb{E}(\mathbb{R}^n)$-structure for the region U of M as the chart relative to which all of the others are described. Any other chart $(U, \bar{x})$ in the $\mathbb{E}(\mathbb{R}^n)$-structure determines a coframe, that is, an equivalence class $[(\bar{x} - \bar{x}(p)) \circ x^{-1}]_p$ defined above in the first part of this appendix. Any two elements of such an equivalence class are related by an element of $O(\mathbb{R}^n)$. Since at each $p \in U$, a set of $O(\mathbb{R}^n)$-related coframes is determined by the $\mathbb{E}(\mathbb{R}^n)$-structure, this structure determines a cross section of $O(\mathbb{R}^n) \backslash \mathscr{L}^*(U)$, that is, a Riemannian structure on U. Evidently, this structure is flat. Conversely, a cross section of $O(\mathbb{R}^n) \backslash \mathscr{L}^*(U)$ for which the Riemann curvature tensor is everywhere zero determines a flat Riemannian structure on U and hence an $\mathbb{E}(\mathbb{R}^n)$-structure for U. Specifically, there exists an adapted coordinate chart (U, x) with respect to which the metric tensor is δ_{ij} throughout U. The $\mathbb{E}(\mathbb{R}^n)$-structure is the set of charts

$$\{(U, a \circ x) \,|\, a \colon \mathbb{R}^n \to \mathbb{R}^n \text{ is Euclidean}\},$$

where the domain of a is of course restricted to $x_+(U)$.

NOTES

* This work was carried out while the authors were participants in a research group at the Zentrum für interdisziplinäre Forschung at the Universität Bielefeld in Bielefeld, Germany, during the academic year 1992–1993.

[1] For another approach to this problem see [23, Schmidt 1979].

[2] 'HILC' stands for *homongeneous isotropic local congruence*.

[3] The parameter, determined a posteriori, is essentially the length of the rods in each class.

[4] Again, the parameter, determined a posteriori, is essentially the length of the rods in each class.

[5] In principle, S need not be linear, however, if the transformation S is very complicated, so is the analysis that is required to recover the family $\mathscr{X}$ from the family $\mathscr{Y}$.

[6] We say 'essentially' because Weyl did not establish theoretic completeness with respect to physical coordinates.

[7] It is important to emphasize that all of the particles used in the measurement procedure must be known to belong to a specific monopole class *before* they are used in the measurement procedure.

REFERENCES

[1] Coleman, R. A. and Korté, H.: 1995, 'A New Semantics for the Epistemology of Geometry I, Modeling Spacetime Structure', this issue.

[2] Coleman, R. A. and Korté, H.: 1980, 'Jet Bundles and Path Structures', *The Journal of Mathematical Physics* **21**(6), 1340–1351.

[3] Coleman, R. A. and Korté, H.: 1981, 'Spacetime *G*-Structures and their Prolongations', *The Journal of Mathematical Physics* **22**(11), 2598–2611.

[4] Coleman, R. A. and Korté, H.: 1982, 'Erratum: Jet Bundles and Path Structures', *The Journal of Mathematical Physics* **23**(2), 345. Erratum of [2].

[5] Coleman, R. A. and Korté, H.: 1982, 'The Status and Meaning of the Laws of Inertia', in *The Proceedings of the Biennial Meeting of the Philosophy of Science Association*, East Lansing, Michigan, pp. 257–274.

[6] Coleman, R. A. and Korté, H.: 1984a, 'Constraints on the Nature of Inertial Motion Arising from the Universality of Free Fall and the Conformal Causal Structure of Spacetime', *The Journal of Mathematical Physics* **25**(12), 3513–3526.

[7] Coleman, R. A. and Korté, H.: 1984b, *A Realist Field Ontology of the Causal-Inertial Structure (the Refutation of Geometric Conventionalism)*. University of Regina Preprint, 192 pages, March 1984. Final enlarged version, entitled *The Philosophical and Mathematical Foundations of Spacetime Theories*, to appear as a volume of *The Synthese Library Series*.

[8] Coleman, R. A. and Korté, H.: 1987, 'Any Physical, Monopole, Equation-of-Motion Structure Uniquely Determines a Projective Inertial Structure and an $(n - 1)$-Force', *The Journal of Mathematical Physics* **28**(7), 1492–1498.

[9] Coleman, R. A. and Korté, H.: 1989, 'All Directing Fields that are Polynomial in the $(n - 1)$-Velocity are Geodesic', *The Journal of Mathematical Physics* **30**(5), 1030–1033.

[10] Coleman, R. A. and Korté, H.: 1990, 'Harmonic Analysis of Directing Fields', *The Journal of Mathematical Physics* **31**(1), 127–130.

[11] Coleman, R. A. and Korté, H.: 1990, 'The Physical Initial Value Problem for the General Theory of Relativity', in Cooperstock, F. I. and Tupper, B. (eds.), *General Relativity and Relativistic Astrophysics*, Singapore, pp. 188–193. World Scientific. Proceedings of the third Canadian Conference on General Relativity and Relativistic Astrophysics held in Victoria.

[12] Coleman, R. A. and Korté, H.: 1991, 'An Empirical, Purely Spatial Criterion for the Planes of *F*-Simultaneity', *Foundations of Physics* **24**(4), 417–437.

[13] Coleman, R. A. and Korté, H.: 1992, 'On Attempts to Rescue the Conventionality Thesis of Distant Simultaneity in STR', *Foundations of Physics Letters* **5**(6), 535–571.

[14] Coleman, R. A. and Korté, H.: 1992, 'The Relation between the Measurement and Cauchy Problems of GTR', in Sato, H. and Nakamura, T. (eds.), *The Sixth Marcel Grossmann Meeting on General Relativity*, World Scientific, pp. 97–119. Printed version of an invited talk presented at the meeting held in Kyoto, Japan, 23–29 June 1991.

[15] Coleman, R. A. and Korté, H.: 1994a, 'Constructive Realism', in Majer, U. and Schmidt, H.-J. (eds.), *Semantical Aspects of Spacetime Theories*, Wissenschaftsverlag, pp. 67–81.

[16] Coleman, R. A. and Korté, H.: 1994b 'A Semantic Analysis of Model and Symmetry Diffeomorphisms in Modern Spacetime Theories', in Majer, U. and Schmidt, H.-J. (eds.), *Semantical Aspects of Spacetime Theories*, Wissenschaftsverlag, Mannheim, pp. 83–94.

[17] Crapo, E.: 1982, 'The Tetrahedral-Octahedral Truss', *Struct. Topol.* **7**, 51–61.

[18] Ehlers, J., Pirani, R. A. E. and Schild, A.: 1972, 'The Geometry of Free Fall and Light Propagation', in L. O' Raifeartaigh, L. O. (ed.), *General Relativity, Papers in Honour of J. L. Synge*, Clarendon Press, Oxford, pp. 63–84.

[19] Guillemin, V. and Sternberg, S.: 1964, 'An Algebraic Model of Transitive Differential Geometry', *Bull. Amer. Math. Soc.* **70**, 16–47.

[20] Guillemin, V. and Sternberg, S.: 1966, 'Deformation Theory of Pseudogroup Structures', *Mem. Amer. Math. Soc.* **64**.

[21] Kretschmann, E.: 1917, 'Über den physikalischen Sinn der Relativitätstheorie', *Annalen der Physik* **53**(16), 576–614.

[22] Kobayashi, S. and Nomizu, K.: 1963, *Foundations of Differential Geometry*, Vol. 1, Interscience, New York.

[23] Schmidt, H.-J.: 1979, *Axiomatic Characterization of Physical Geometry*, Vol. 111 of *Lecture Notes in Physics*, Springer-Verlag, Heidelberg. Edited by J. Ehlers et al.

[24] Singer, I. M. and Sternberg, S.: 1965, 'The Infinite Groups of Lie and Cartan', *Ann. Inst. Fourier. I.* **15**, 1–114.

[25] Weyl, H.: 1921, 'Zur Infinitesimalgeometrie: Einordnung der projektiven und konformen Auffassung', *Nachr. Königl. Ges. Wiss. Göttingen, Math.-phys. Kl.*, pp. 99–112. Reprinted in [26].

[26] Weyl, H.: 1968, *Gesammelte Abhandlungen*, Vols. 1–4, Springer Verlag, Berlin. Edited by K. Chandrasekharan.

[27] Whiteley, W.: 1982, 'Motions of Trusses and Bipartite Frameworks', *Struct. Topol.* **7**, 61–68.

University of Regina
Math. Physics and Philosophy of Science
Regina
Saskatchewan S4S OA2
Canada

HEINZ-JÜRGEN SCHMIDT

A MINIMAL INTERPRETATION OF GENERAL RELATIVISTIC SPACETIME GEOMETRY

1. INTRODUCTION

A mathematical theory usually admits of various equivalent formulations. It is not only possible to choose different sets of axioms but also the decision of what is considered as a basic concept and what as a derived one is, to some extent, optional.

It is well known that, for example, a *topology* can be characterized in terms of *open sets* or, alternatively, in terms of *neighbourhoods* of points. Other examples are given by synthetic versus analytic formulations of, say, projective geometry.

Given the manifold of different formulation of "one" theory, it is a legitimate mathematical objective to choose the basic concepts as simply as possible. Of course, 'simple' is only vaguely defined, it may happen that two mathematicians disagree about that question. Moreover, the simplicity of the basic concepts possibly has to be purchased with the abundance and complexity of the corresponding axioms.

For example, Robb (1936) proved that Minkowski spacetime can be characterized by means of a binary relation of causal order between events but he has to invoke dozens of axioms. In the case of Euclidean geometry the situation is more balanced: Tarski gave an exposition of this theory based on a ternary relation of betweenness and a quaternary congruence relation (Tarski, 1959). Alternatively, I have shown[1] that as basic concepts of Euclidean geometry a set of "regions" together with two binary relations of "inclusion" and "congruence" would suffice. In both cases the number of axioms can be kept small.

In most cases it will be a simpler task to show that it is *possible* to give an equivalent formulation T_2 of a given theory T_1 using certain basic concepts than to explicitly formulate T_2. For the first task it suffices to show how the basic concepts B_1 of T_1, can be "redefined" within T_1 in terms of the basic concepts of T_2, denoted by B_2. What is this to mean? Using the model-theoretic notion of 'definition', we may more exactly describe the first task as follows:

Given two T_1-models which are B_2-isomorphic, show that they are also B_1-isomorphic!

For the above example of Robb's characterization of Minkowski-spacetime this amounts to proving that any bijection $f: M \rightarrow M'$ of Minkowski spaces which preserves the causal order also preserves the affine and metric structure (up to a constant factor). This is essentially the

71

Erkenntnis **42**: 191–202, 1995.
© 1995 *Kluwer Academic Publishers*.

Alexandrov–Zeeman theorem (Alexandrov, 1967; Zeeman, 1964) which, although published later, follows as a corollary of Robb's result.

In this paper I will only solve the first task to show how the relevant structures of pseudo-Riemannian spacetime can be re-defined in terms of the ternary betweenness relation β. It would be much more difficult to solve the second task, i.e. to formulate axioms for β and to prove a suitable representation theorem, even if the result (Ehlers et al., 1972) of Ehlers, Pirani, and Schild (EPS) and related work could be used.

Up to now we have only considered equivalent formulations of *mathematical* theories *T*. Turning to *physical* theories we have to take into account that some or all mathematical terms of *T* will have some physical interpretation, whatever this means. It is an old idea underlying attempts to give "physical axiomatizations" of theories that among the different equivalent formulations of *T* some are easier to interprete because their basic concepts are closer to physical experience. For example EPS wrote (Ehlers et al., 1972):

... we attempt to derive the conformal, projective, affine, and metric structures of space time from some qualitative (incidence and differential topological) properties of the phenomena of light propagation and free fall that are strongly suggested by experience. Not only the measurement of length but also that of time then appears as a derived operation.

The conceptual reduction which is achieved by an axiomatization of this kind has to be considered with some care. For example, the integral $\int ds$ along some time-like curve is physically interpreted as the lapse of time measured by an ideal clock. Actually, this interpretation is the basis of some experimental tests of the general theory of relativity (GTR), e.g. the Hafele–Keating experiment (Hafele and Keating, 1972) using atomic clocks as approximate realizations of ideal clocks. On the other hand, EPS define the metric and hence $\int ds$ in terms of the projective and conformal structure, which represent the world-lines of freely falling test particles and (test) light-rays. It follows from the EPS formulation that ideal clocks can be constructed from freely falling particles and light-rays (for one of the different possible constructions see below) and that ideal clocks measure $\int ds$, but it would be a contingent fact that e.g. atomic clocks practically agree with ideal clocks. Hence a negative outcome of the Hafele–Keating experiment (including all fictitious attempts to build improved clocks) would not falsify EPS–GTR but only EPS–GTR plus the extra hypothesis that atomic clocks are good approximations of ideal clocks. Note that the situation is different for the "usual" interpretation of GTR, where the extra hypothesis is needed to give a physical interpretation of $\int ds$. Hence the EPS-analysis reveals the fact that the extra hypothesis is, strictly speaking, not necessary for interpreting GTR, although it is fulfilled and important for many applications. From this point of view the semantical poverty of the EPS formulation is not a draw-back, but rather a virtue, as long as it is not pretended that one has reconstructed the full-fledged GTR.

Analogous arguments apply to an even more austere formulation like the one indicated in the present paper. It provides a minimal interpretation of GTR (where 'minimal' is not meant in a strict mathematical sense) which may be interesting for the conceptual analysis but is not intended to account for the variety of "real life" applications of GTR.

As mentioned above I will exploit the EPS result. A closer inspection (Meister, 1991) shows that their basic concepts can be written as $(M; \mathcal{P}, \mathcal{L}, \mathcal{U})$, where M is the set of *events*, $\mathcal{P}$ the set of *particles*, $\mathcal{L}$ the set of *light-rays*, and a particle or light-ray is taken as a subset of M. Additionally, the structure $\mathcal{U}$ has to be considered as basic, although (incorrectly) hidden by EPS in the form of an existential statement in their axiom D_1. $\mathcal{U}$ can be taken as a collection of maximal atlases, which make each particle into a 1-dimensional C^3-manifold. Physically, $\mathcal{U}$ can be thought of as a collection of "pre-clocks" which measure not necessarily proper time but a C^3-function of proper time along particles. Pre-clocks are essential for the construction of radar charts in EPS which constitute an operational realization of the differentiable structure of M.

If one wants to show that spacetime can be characterized as a pair (M, β) it suffices to show that $\mathcal{P}$, $\mathcal{L}$ and $\mathcal{U}$ can be defined in terms of β within any pseudo-Riemannian spacetime geometry. The plan of showing this goes as follows. First we will define *particles* by means of β, together with the relation of *time-like separation* of events and a topology on M. Next, we will define $\mathcal{L}$ or, equivalently, *light-like separation* as the limit of time-like separation. These steps are rather elementary and will be sketched in Section 2. In Section 3 we will define $\mathcal{U}$ in terms of $\mathcal{P}$ and $\mathcal{L}$ and hence indirectly in terms of β. To this end we analyze the construction of clocks according to Castagnino (1968). To my knowledge this is the only construction which does not presuppose the differentiable structure of spacetime,[2] which would of course be circular for the present purpose. In Minkowski spacetime. Castagnino's clock measures proper time by virtue of Desargues' theorem. In GTR spacetime Desargues' theorem holds locally up to corrections of third order of the size of the considered region (Heilig and Pfister, 1990) and hence it is plausible that Castagnino's clock will work here too. An explicit proof of this will not be given in this paper but will be published elsewhere. Note that the proof would yield more than we need since for $\mathcal{U}$ we need not define clocks but only pre-clocks. Thus it makes sense to ask under which weaker conditions Castagnino's construction still works and will yield not clocks but pre-clocks. It turns out that a possible weakening consists of taking $\mathcal{P}$ as the set of time-like solutions of a second order equation of motion, not necessarily a geodesic one. Of course, $\mathcal{P}$ has still to be compatible with the conformal structure of spacetime in the sense that the light cone with vertex A is the boundary of the union of all particles through A.

This possibility might be of interest for those who prefer an axiomatization of GTR spacetime which does not start with a single $\mathscr{P}$ but with a collection of $\mathscr{P}$'s and provide criteria which distinguish the geodesic equation of motion $\mathscr{P}$.[3] Our approach shows that this would also be possible without presupposing $\mathscr{U}$. It may even be conjectured that the criterion for geodesicity is equivalent to the condition that the readings of pre-clocks are independent of the auxiliary points needed for the present construction.

It should be noted that a mathematical result saying that the relation β already determines the C^∞-structure of M is not the strongest possible one. It is known[4] that under certain conditions the conformal structure of spacetime already determines the differentiable structure. The proof of that result is, however, not constructive in contrast to the present approach.

2. PARTICLES, TOPOLOGY AND LIGHT-RAYS

In flat spacetime the *particle* p containing two given different events A, B can be defined as the minimal subset of M containing A and B and being closed with respect to β.[5] In curved spacetime *caustics* may occur, i.e. different geodesics containing the same events A, B. In that case, p will not be closed with respect to β and will only be unique if A and B are close enough. Moreover, the complication of closed particles (geodesic loops) has to be taken into account.

It seems to be most convenient to restrict the β-relation to the case where the events are "close". Recall[6] that for a given pseudo-Riemannian space-time $(M; \mathscr{D}, g)$ a *convex normal neighbourhood* $\mathscr{M}_p$ is defined as a neighbourhood of $p \in M$ such that exp: $U \subset T_p M \to \mathscr{M}_p$ is a diffeomorphism and for any A, C with $A \neq C$ there is a unique geodesic inside $\mathscr{M}_p$ joining A and C. Let M be covered by a countable family $(\mathscr{M}_i)_{i \in I}$ of convex normal neighbourhoods. Then we define:

DEFINITION 1. Let $A, B, C \in M$, then $\beta(A, B, C)$ holds iff $A \neq C$ and there exists some $i \in I$ such that $A, B, C \in \mathscr{M}_i$ and the unique geodesic joining A and C is time-like and has a parametrization $\gamma: \mathbb{R} \to M$ such that for three real numbers $a < b < c$ we have $\gamma(a) = A$, $\gamma(b) = B$, $\gamma(c) = C$.

With this restriction of betweenness we can define a sensible concept of *local particles* (or simply *particles*) by

$$(1) \qquad p(A, C) \overset{\text{def}}{=} \{B \in M \,|\, \beta(A, B, C)\} = p(C, A).$$

A, C are said to be time-like related if $p(A, C)$ is not empty. In that case the class of local particles which are subsets of $p(A, C)$ form a basis of open sets for a topology on $p(A, C)$. *Global particles* will be defined as

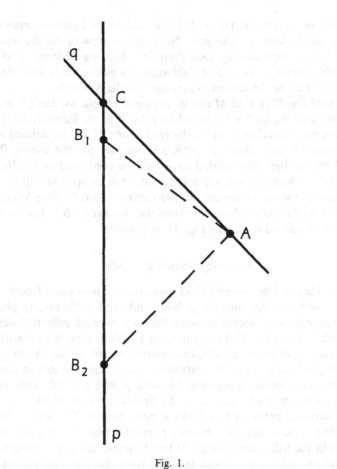

Fig. 1.

unions of nested local particles $\cup_{i=0}^{N} p(A_i, A_{i+2})$, where $\beta(A_i, A_{i+1}, A_{i+2})$ holds for $i = 0, \ldots, N$. *Maximal particles* are defined with respect to the obvious notion of inclusion of global particles.

Light-rays are determined by light cones, hence by the relation between events of being light-like related. The details of this determination involve the differentiable structure $\mathscr{D}$ of M and may be omitted here since for the construction of (pre-)clocks we only need light-like relatedness. This in turn can be defined as the limit of time-like relatedness. Consider a local particle p and an event A near p but $A \notin p$. Then there exist exactly two events B_1, B_2 with the following properties[7] (see Fig. 1):

- $B_1, B_2 \in p$,
- no particle contains A and B_1 (resp. A and B_2),
- in every p-neighbourhood of B_1 (resp. B_2) there is an event $C \in p$ and a particle q with $A, C \in q$.

In this case A and B_1 (resp. B_2) are said to be *light-like related* and (A, p) is said to form a *radar pair*. Note that in view of the discussion in Section 1 it is not necessary that there is a light-signal from A to B or conversely, although the use of light-signals would considerable simplify the construction of Gedanken experiments realizing clocks.

The usual topology $\mathcal{O}$ of M can be regained as follows: Let (A, p) be a radar pair and $B_1, B_2 \in p$ be light-like related to A. Further, let U_1, U_2 be p-open neighbourhoods of B_1, B_2 resp. Then U will be defined as the set of events A' such that (A', p) is a radar pair and the events B'_1, B'_2 on p which are light-like related to A' will be contained in U_1, U_2 resp. The set of U's such defined will form a sub-basis of open sets in $\mathcal{O}$.

For later purposes we remark that spacetime is locally time-orientable, i.e. locally a continuous choice between the two events B_1, B_2 on p which are light-like related to A (see Fig. 1), is possible.

3. CASTAGNINO'S CLOCK

The basic idea of Castagnino's clock construction is to use a freely falling rigid rod with parallel mirrors at both ends and a "bouncing photon" which is repeatedly reflected between the mirrors, each reflection defining one "click" of the clock. The remaining problem then is to construct a rigid rod or, more geometrically, to construct a curve p' parallel to a given time-like geodesic p. In GTR spacetime, this is only possible in the limit of infinitesimal spatial separation between p and p'. This limit is also necessary for arbitrarily increasing the precision of the clock, especially for the duration between two clicks approaching 0. The second idea of Castagnino is, in order to construct parallels, that of using Desargues' theorem in the following version: If two triangles are perspective from a point and two sides are pairwise parallel, then also the third pair of sides will be parallel. Moreover, in a spacetime plane two light-rays going into the same direction will be parallel, so these can be used in the Desargues figure, see Fig. 2.

Of course, all this is only approximately satisfied in GTR spacetime. The resulting figures for the construction of Castagnino's clock can be simplified a bit if some of the Desargues auxiliary light-like lines are chosen to coincide with some world lines of the bouncing photon.

We will now describe the construction of Castagnino's clock in detail.[8] The only construction principle we need is the following:

DEFINITION 2. Given a particle r and a near-by event $A \notin r$ then $B = \mathcal{E}(A, r)$ is the unique event on r which is "future-light-like" related to A. By "future-light-like" we refer to one of the two local time orientations, no matter which one. The operator $\mathcal{E}$ will also be used with the syntax $B = \mathcal{E}(A, C, D)$ which is equivalent to $B = \mathcal{E}(A, r)$ if C, D are time-like related and r is the unique particle containing C and D.[9]

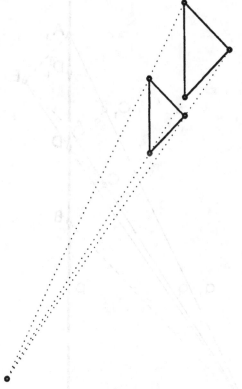

Fig. 2. Desargues' figure.

Let be given a global particle p and two events A and Z lying on p. A chosen local time-orientation can be extended to a neighbourhood of $p(A, Z)$ and with respect to this time-orientation A is the "begin" of the clock and Z the "end". Further, we choose an auxiliary particle q with $A \in q$ and an auxiliary event $Q \in q$ which is "earlier" than A in the above sense.

Next we construct 7 events according to the following definitions, see Fig. 3:

(2) $B \overset{\text{def}}{=} \mathscr{E}(Q, p)$

(3) $C \overset{\text{def}}{=} \mathscr{E}(B, q)$

(4) $D \overset{\text{def}}{=} \mathscr{E}(C, p)$

(5) $E \overset{\text{def}}{=} \mathscr{E}(A, Q, D)$

(6) $A_1 \overset{\text{def}}{=} \mathscr{E}(E, p)$

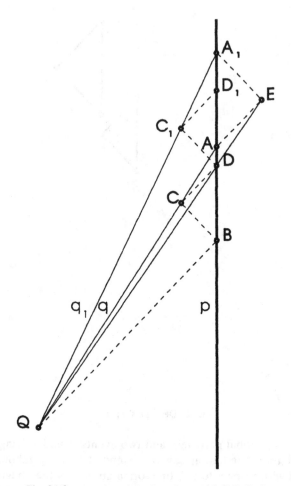

Fig. 3. Construction of the first three "ticks" B, D, D_1.

$$
(7) \qquad C_1 \overset{\text{def}}{=} \mathscr{E}(D, Q, A_1)
$$

$$
(8) \qquad D_1 \overset{\text{def}}{=} \mathscr{E}(C_1, p)
$$

We note that, if the construction would be performed in Minkowski spacetime, the triangles A, A_1, E and C, C_1, D would be perspective from Q and the light-rays E, A_1 and D, C_1 be parallel, likewise A, E and C, D. Hence, according to Desargues' theorem, the lines through D, D_1 and C, C_1 would also be parallel. The first line is, of course, p but the second one does not explicitly occur in our construction. The path of the "bouncing photon" will be B, C, D, C_1, D_1, hence B, D, D_1 will be the first three clicks of our clock. It is obvious how the next click D_2 has to be defined:

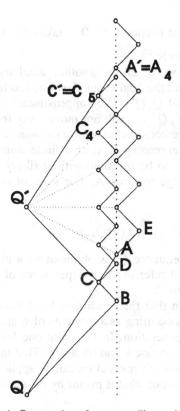

Fig. 4. Construction of a new auxiliary point Q'.

Define the auxiliary events E_1, A_2, C_2 and then D_2 analogously to the above definitions, if C and D are replaced by C_1 and D_1.

It is tempting to describe the rest of the construction simply by "and so on". But the uniqueness of the particles containing two events, for example Q, A_n, presupposes that these events should remain close. More important, the calculations which are to show that Castagnino's clock works presuppose that the auxiliary particles are confined to *arbitrarily* small regions. Hence the auxiliary event Q has to be carried along after a finite number of clicks. This can be done as follows: Since the distance between the A_i is greater than the distance between the C_i it will happen after a finite number of clicks that A_i, C_{i+1} will be time-like related. Then a new auxiliary event Q' can be defined by (see Fig. 4)

$$(9) \qquad Q' \stackrel{\text{def}}{=} \mathscr{E}(B, A_i, C_{i+1}).$$

The above construction of subsequent clicks will then be repeated such that the role of the first events A, C, D is taken by $A' \overset{\text{def}}{=} A_i$, $C' \overset{\text{def}}{=} C_{i+1}$, and $D' \overset{\text{def}}{=} D_i$, respectively.[10]

After a finite number of clicks, another auxiliary point Q'' will be defined, and so on, until the final event Z will be reached. Again, Desargues' theorem shows that Q, Q' will be approximately parallel to p, hence the auxiliary events $Q', Q'', \ldots$ will not move away from p.

Up to now, we have constructed an *approximate clock* starting somewhere before A and exceeding Z after a finite number of clicks. This whole construction has to be repeated with auxiliary events $Q_k \in q$, $k = 1, 2, \ldots$ which converge towards A. For sake of definiteness, we will define

$$(10) \qquad Q_0 \overset{\text{def}}{=} Q, \quad Q_{k+1} \overset{\text{def}}{=} \mathscr{E}(\mathscr{E}(Q_k, p), q).$$

This means that the sequence Q_k is obtained by a photon bouncing between q and p. We will refer to the k-dependence of the construction as to different "generations".

It could well happen that the conditions for applying the $\mathscr{E}$-operator, namely that the events occuring as arguments of $\mathscr{E}$ are close, are violated in the course of one generation. In that case one has to go back to the auxiliary event Q_k, to increase k and try again. That there is some k_0 such that for all $k > k_0$ the whole construction can be applied without problems until Z is reached, is part of what is meant by the statement 'Castagnino's clock works'.

Let X be an arbitrary event on p "between"[11] A and Z and denote by $i = N(X, k)$ the number of clicks in the kth clock construction such that $\beta(D_i, X, D_{i+1})$ holds. Then we have a real parameter λ on p defined by

$$(11) \qquad \lambda(X) \overset{\text{def}}{=} \lim_{k \to \infty} \frac{N(X, k)}{N(Z, k)}.$$

That this limit exists is part of the correct functioning of Castagnino's clock and the central statement to be proven in a complete treatment of the subject.

The next step would be to extend the above construction to get "coordinate grids" and, in the limit, coordinate charts which give rise to smooth coordinate transformations. The differentiability properties should not be postulated ad hoc as it is usually done but should be obtained in a natural way. Of course, this is a very difficult task.

NOTES

[1] See Schmidt (1979) and the Appendix A in Schmidt (1994).
[2] See also the discussion in Perlick (1994).

[3] This especially applies to the approach of Coleman and Korté, see Coleman and Korté (1982, 1987) and Coleman and Schmidt (1994).

[4] See Hawking (1966); Hawking et al. (1976) and Malament (1977).

[5] I.e. $X, Y \in p$ and $\beta(\pi(X, Y, Z))$, where π is some permutation, implies $Z \in p$.

[6] See, for example, Theorem 6.2 in Helgason (1962).

[7] According to our definition of 'local particle', it would be more exact to replace p in the items by a some local extension p', which will be again a local particle. In favour of readability I will not be too pedantic about this point. Thus a local extension of particles will be understood if necessary.

[8] Although I will keep the term 'Castagnino's clock' it should be clear that I am responsible for its detailed design and corresponding possible errors.

[9] Again, local extension is understood.

[10] For later purposes we additionally continue the counting of the D-clicks, $D_{i+1} = D'_i$ and so on.

[11] $\beta(A, X, Z)$ need not hold because of our restricted definition of β.

REFERENCES

Alexandrov, A. D.: 1967, 'A Contribution to Chronometry', *Can. J. Math.* **19**, 1119–28.

Castagnino, M.: 1968, 'Some Remarks on the Marzke–Wheeler Method of Measurement', *Nuovo Cimento B* **54**, 149.

Coleman, R. A. and Korté, H.: 1982, Let Bundles and Path Structures', *J. Math. Phys.* **21**(6), 1340–1351. *Erratum* **23**, 345.

Coleman, R. A. and Korté, H.: 1987, 'Any Physical, Monopole, Equation of Motion Structure Uniquely Determines a Projective Inertial Structure and an $(n-1)$-Force', *J. Meth. Phys.* **28**(7): 1492–1498.

Coleman, R. and Schmidt, H.-J.: To appear, 'A Geometric Formulation of the Equivalence Principle', *J. Math. Phys.*

Ehlers, J., Pirani, F. A. E., and Schild, A.: 1972, 'The Geometry of Free Fall and Light Propagation', in L. O'Raifeartaigh (ed.), *General Relativity, Papers in Honour of J. L. Synge*, Clarendon Press, Oxford, pp. 63–84.

Hafele, J. C. and Keating, R.: 1972, 'Around-the-World Atomic Clocks', *Science* **177**, 166–170.

Hawking, S. W.: 1966, *Singularities and the Geometry of Space-Time*. Adams Prize Essay, unpublished.

Hawking, S. W., King, A. R., and McCarthy, P. J.: 1976, 'A New Topology for Curved Space-Time which Incorporates the Causal, Differential, and Conformal Structures', *J. Math. Phys.* **17**(2): 174–181.

Heilig, U. and Pfister, H.: 1990, 'Characterization of Free Fall Paths by a Global or Local Desargues Property', *J. Geom. Phys.* **7**, 419.

Helgason, S.: 1962, *Differential Geometry and Symmetric Spaces*, Academic Press, New York.

Malament, D. B.: 1977, 'The Class of Continuous Timelike Curves Determines the Topology of Spacetime', *J. Math. Phys.* **18**(7), 1399–1404.

Meister, R.: 1991, *Eine Neuformulierung der EPS-Axiomatik*, Diploma Thesis, Universität-GHS, Paderborn.

Perlick, V.: 1994, 'Characterization of Standard Clocks in General Relativity', in U. Majer and H.-J. Schmidt (eds.), *Semantical Aspects of Spacetime Theories*, BI-Wissenschaftsverlag, Mannheim, pp. 169–180.

Robb, A. A.: 1936, *Geometry of Time and Space*, The University Press, Cambridge. Revised edition of the 1914 book.

Schmidt, H.-J.: 1979, *Axiomatic Characterization of Physical Geometry*, Vol. 111 of *Lecture Notes in Physics*, Springer, Berlin.

Schmidt, H.-J.: 1994, 'Holistic versus Hierarchic Concepts of Theories', in U. Majer and H.-J. Schmidt (eds), *Semantical Aspects of Spacetime Theories*, BI-Wissenschaftsverlag, Mannheim.

Tarski, A.: 1959, 'What is Elementary Geometry?', in L. Henkin, P. Suppes, and A. Tarski (eds), *The Axiomatic Method with Special Reference to Geometry and Physics*, North-Holland, Amsterdam.

Zeeman, E. C.: 1964, 'Causality Implies the Lorentz Group', *J. Math. Phys.* 5, 490–493.

Department of Physics
University of Osnabrück
Barbarastraße 7
D-49069 Osnabrück
Germany

FELIX MÜHLHÖLZER

SCIENCE WITHOUT REFERENCE?

1. SEMANTICAL ASPECTS OF SCIENTIFIC THEORIES

One of the central ideas of modern science is an idea of maturity: Just as a child, in order to grow up, has to grow out of its childlike fantasies when dealing with the world, so the scientist when dealing with nature has to replace his egocentric and anthropocentric everyday ideas with ways of thinking which are more adequate to objective reality. This may sound like a religious or even mystical attitude, and it has in fact been expressed as such with impressive pathos by scientists themselves, for example by Albert Einstein.[1] In science, however, this attitude takes a specific shape which distinguishes it rather clearly from the genuinely religious or mystical attitude. Whereas religion or mysticism is mostly directed toward the 'inner world' and strives for immediate access to what this inner world reveals, science mostly turns to the 'outer world' which it tries to uncover through theoretical constructions. The scientific attitude is to check the theoretical constructions by means of outward directed experience and always to be ready to be corrected by this experience. This is also the really essential and important tenet of what is called "empiricism".

Strangely enough, in Logical Empiricism, i.e., in the variety of empiricism started by the members of the Vienna Circle,[2] this tenet has been transformed into a 'semantical' one; "semantical" in the general sense of a 'theory of (cognitive) meaning'.[3] Roughly speaking, the idea was simply to identify just those experiences which we deem to be relevant when we search for the truth value of a sentence with the 'meaning' or the 'sense' of that sentence. This is the notorious Verifiability Theory of Meaning.[4] It is even upheld by someone like Quine who, in other respects, is one of the most vehement critics of Logical Empiricism. See, for example, what Quine says in a reply to Roger Gibson: "I find [verificationism] attractive. The statement of verificationism [. . .] is that "evidence for the truth of a sentence is identical with the meaning of the sentence"; and I submit that if sentences in general had meanings, their meanings would be just that".[5]

Of course, in Quine's work this idea is, in a sense, reduced to absurdity by means of his holism. The passage just quoted continues as follows: "It is only holism itself that tells us that in general they [= sentences] do not have them [= meanings in the verificationist sense]." According to Quinean holism, only whole theories may have well-defined empirical test situations, but in general this is not the case for single sentences. In the

Erkenntnis **42**: 203–222, 1995.

verificationist spirit this means that only whole theories, but in general not single sentences, may have meaning. This position is called "meaning holism".[6]

The really important part of Quinean meaning holism is its negative part: Single sentences, i.e., sentences considered in isolation from other sentences, do not in general have meaning. Its positive part – whole theories do have meaning – however, is rather dubious. For Quine, a theory essentially is nothing but a class of sentences, and the relation between theory and experience is reduced to the logical relations between the sentences of the theory and true observation sentences – or, more precisely, true 'observation categoricals'. A Quinean observation categorical is a sentence like "Where there is smoke there is fire", compounded of two observation sentences ("Here is smoke", "Here is fire") by means of the "Whenever this, that"-construction.[7] Quine then simply defines the 'meaning' – or the 'empirical content' – of a theory as the class of observation categoricals logically implied by the theory.[8] This view, however, has a strong air of fiction. The reason lies in the ubiquity of what scientists regard as 'disturbing factors'. Observation categoricals are sentences of strict generality which, besides being wrong in most cases, in general are *not* logically implied by our existing theories. This has been conspicuously worked out by Hempel.[9] In general, our theories deal with undisturbed, ideal situations. In most cases their applications to real situations are subject to certain *provisos* to the effect that no disturbing factors are present. Seen from the observational level, these disturbing factors are unsurveyable and cannot be specified in advance. Precisely such specifications, however, would be necessary if our theories were able to *logically* imply observation categoricals. Thus Quine's account of 'empirical content' is based on a strongly idealized picture of the relation between theory and evidence.[10]

Wittgenstein would say that Quine's notion of a scientific theory is nothing but a philosopher's *Luftgebäude* (a building made of air).[11] As soon as one uses the word "theory" in the manner of the practicing scientist, the relation between theory and experience no longer appears as an affair which can be described by a short formula like "logical implication of observation categoricals". Theories in the scientists' sense reveal themselves to be very flexible entities, not sharply defined sentence-buildings, and the answer to the question of whether a certain experimental outcome contradicts a theory depends on the skill with which theoreticians and experimenters make use of the theory. A really good reason to consider a theory falsified will be given only if a better theory is available which explains why the adherents of the old theory were unsuccessful. In other words, an isolated theory has no well-defined range of falsifying observable situations. Thus, Quine's holistic version of verificationism may be proven untenable by a line of argumentation resembling the argumentation which Quine himself uses to refute sentence-oriented verificationism.

Seen in this way, verificationism does not appear attractive at all, and in particular the verificationists' hierarchical picture, in which meaning is flowing from the basis of observation sentences 'upward' into the more and more theoretical domains, proves to be dubious. This picture, of course, has also been attacked from quite a different angle by people like Hanson, Feyerabend and Kuhn, who stress what they call the "theory-ladenness" of observation sentences.[12] From their point of view meaning also flows 'downward' from the more theoretical to the more observational parts of our language, with the effect that the theoretical-observational dichotomy itself begins to dissolve.[13] The lesson of all this should be, I think, that meaning may flow in any direction whatsoever; or even better: that one should be careful with metaphors of this kind.

The failure, or at least the only slight usefulness, of verificationism has evoked quite different reactions. Quine himself has been driven into some sort of meaning nihilism; others have been looking for alternative conceptions of meaning. One may ask, then, what these alternative conceptions might achieve within the philosophy of *science*. The suspicion may arise that in philosophy of science there is no room left for any sensible application of theories of meaning or for the philosophy of language in general. It seems to me that much of contemporary philosophy of science is dominated by just this suspicion.

But not all of it. An important exception is, for example, the work of Thomas Kuhn. One central notion of Kuhn's philosophy of science, his notion of *incommensurability*, is essentially a semantical one. To be incommensurable means to lack a common measure, and Kuhn's idea is that the common measure which incommensurable theories are lacking is simply the meaning of certain central terms. For example, with respect to the transition from Newtonian to relativistic physics – the *Newton–Einstein transition*, for short – which Kuhn calls "a prototype for revolutionary reorientations in the sciences" and which, of course, is also a prototype for incommensurability, Kuhn diagnoses a "need to change the meaning of established and familiar concepts", like the concepts of space, time and mass. Of the latter ones Kuhn says: "the physical referents of these Einsteinian concepts are by no means identical with those of the Newtonian concepts that bear the same name".[14] In the last statement, Kuhn is not concerned with 'meaning' in general but, more specifically, with 'reference'. So Kuhn not only claims that in the Newton–Einstein transition the meaning of the terms "space", "time" and "mass" has changed (in whatever sense of "meaning"), but, much more strongly, that even their reference has changed. Consequently, he talks about the *Newtonian* mass in contrast to the *Einsteinian* one.[15]

In the remainder of this paper, in investigating semantical aspects of scientific theories, I will confine myself mainly to the Newton–Einstein transition as my illustrating example. Furthermore, I want to plead for abandoning the use of the notion of meaning in the sense of 'intension', in contrast to "meaning" in the sense of 'extension'. Roughly speaking, the intension of a word *w* should comprise just that information about the objects referred to by *w* which is 'constitutive of the meaning' of *w*. In the case of the word "bachelor", for example, the information that bachelors are male and unmarried would presumably count as meaning-constitutive; information of the kind that a certain percentage of bachelors live alone would not count as meaning-constitutive. This distinction between meaning-constitutive and non-meaning-constitutive information, however, is a very vague one, and to-date nobody has succeeded in clarifying it in a theoretically satisfying and non-arbitrary way. The word "bachelor" may be an exception; in the case of other words, and especially of the philosophically and scientifically interesting words, the distinction is beset with serious problems, and to my mind the present state of the art in philosophy of language (not to mention linguistics) suggests that these problems are insuperable.[16] So at present one should try to get along without the notion of meaning in the sense of intension.[17]

In what follows, when mentioning the term "meaning" at all, I will mention it only in this dubious sense. Of course, Thomas Kuhn, too, could not ignore the bad shape in which the notion of meaning finds itself. Consequently, in his more recent work, when discussing incommensurability, he shrank from making himself too dependent on it[18] and he re-placed it by the notion of translation. Now, incommensurability was to be untranslatability.[19] But in his most recent work, he also shrinks from making this move[20] and offers us the notion of 'lexical structure' as the decisive term for analysing incommensurability. If I understand him correctly, a lexical structure is a structure of conceptual dependencies which is set down in, as it were, our mental dictionary, i.e., in some sort of dictionary which is to be found in our mind/brain. And incommensurability is now seen as incompatibility of lexical structure.[21] However, it seems to me that with this last move Kuhn again runs the risk of becoming too dependent on the notion of meaning in all its dubiousness, and at present I do not see how he may escape that. I suspect that he is on the wrong track.

When pleading for dispensing with the notion of meaning I do not plead, of course, for an abandonment of the notion of *reference*. Apparently it is an essential aim of our scientific discourse – and also of our non-scientific discourse – to refer to specific objects about which we want to make assertions. This seems to be something quite elementary and quite obvious. Let me add, however, that, when talking about 'reference to objects', I will use the notion of an object in a very general sense such that, e.g., abstract entities, like numbers or functions – as, e.g., the mass function referred to by the term "mass" – are also to count as objects. In the philosophical tradition, for example in Kant, the notion of an object is very often understood in a narrower sense. In Kant's theoretical philosophy "object" is very often synonymous with "object of experience"; for Kant, then, "object" is a notion of transcendental logic. In the general sense in which I will use it, "object" may rather appear as a notion of formal logic. It is this notion which is found in the work of Frege and is more or less taken for granted by people like Carnap or Quine. For them, to refer to objects just is to use the linguistic devices of singular terms, predication, identity, and quantification to make serious statements.[22]

One essential function of the notion of reference then lies in the fact that with its help we describe the contribution of words to the truth values of sentences. The sentence "Socrates is wise" is true if and only if the object referred to by the name "Socrates" is an element of the class of objects referred to by the predicate "is wise". Elementary connections of this kind constitute the essential formal property of reference.[23] Furthermore, the identity predicate seems to be important for reference. The reason for that has to do with *objectivity*.[24] The objects which we refer to should confront us as objective in the sense that they can be represented in different ways and from different perspectives. That is, *one and the same object* should be capable of being represented in different ways and from different perspectives. Without this possibility our assertions about objects could not have any content that is independent of the context of utterance and, in particular, the utterer himself.

With that we address an idea of objectivity which is well-known from Kant and which, of course, functions as an important norm in the natural sciences. It requires that, at least within one and the same theoretical framework, questions about the identity or non-identity of objects possess definite answers, at least in principle. The question arises, however, of what should be said in the case of different theoretical frameworks, for example in the case of two successive theories that are divided by what Kuhn and others call a "scientific revolution". What should we say, for example, in the case of Niels Bohr's early theory of the atom, on the one hand, and quantum mechanics, on the other?[25] Both theories contain the word "electron"; however, they make very different assertions about their electrons. Do they nevertheless refer to *the same* objects with this word? In this case there is no identity predicate building a bridge between the different ways of using of the word "electron" and the question of identity appears to be rather awkward. Is it also possible in this case to claim identity – or better: *constancy* – of reference? And are there good *reasons* to do so, analogous, say, to the objectivity reasons mentioned above? It is questions like these which give the Kuhnian topics of "scientific revolution" and "incommensurability" their philosophical explosiveness.

Notoriously, in general Kuhn's response to these questions is negative. According to his view of the development of science, scientific revolutions bring along radical changes of reference, to the effect that the scientists before and after a revolution are talking about different worlds (or even 'live' in different worlds).[26] Unclear as this thesis is, with it Kuhn at least succeeds in urging one to think about – and to work out and justify – another, less dramatic view which may let us see more continuity in the development of science.

2. INTERPRETATIVE STORIES

Thomas Kuhn, when claiming changes of reference for important terms of our scientific theories, very often contradicts scientific practice which, among other things, is an interpretative practice. From the viewpoint of their best theories scientists interpret utterances of former scientists, and by doing so they presuppose as far as possible, I think, that reference has not changed. If a physicist of our times looks at Niels Bohr's early theory of the atom he will presuppose that Bohr's term "electron" refers to the same entities as his own term "electron". (Whether these entities should be called "particles" is another question.) Apparently, Bohr himself made this presupposition of sameness of reference, when using the same word "electron" in his early as well as in his later theoretical endeavours.

To be sure, Bohr's early statements were different from his later ones. This difference, however, may be explained without postulating a difference in the extensions of the words the statements are compounded of. One may simply say that the later Bohr changed his mind about just those entities he already had talked about earlier, and one can tell stories about how and why he changed his mind. At this point one may bring in the so-called *causal theory of reference*. The actual importance of the causal theory of reference can be seen in its endeavour to tell interpretative stories of exactly this sort. Seen in this way, the causal theory of reference is nothing but a theory of the constancy of reference over time. It does not attempt to say, then, what reference *is*; rather, it is drawing a picture that shows how the interplay of human action and the human environment makes us assert that reference has not changed. This interplay is a causal one and that is why the theory is justly called a 'causal' theory of reference.

Let me give two examples of interpretative stories of this kind.[27] They concern the notions of simultaneity and mass with regard to the Newton–Einstein transition. Consider, first, simultaneity. What can we say, from the point of view of (special) relativity theory, about the reference of the term "simultaneous" as used by physicists adhering to Newtonian physics ("Newtonian physicists", for short)? Take, for example, a situation which corresponds to the following simple method of determining the simultaneity of events:[28] Consider two identical guns positioned at the midpoint between the points A and B of an inertial frame I. Suppose that cannon balls, simultaneously fired from the guns toward A and B, actually hit A and B. Then, these two hitting events are simultaneous. This is true, from the point of view of special relativity theory, relative to any inertial frame

I, and it would be accepted also by a Newtonian physicist. The Newtonian physicist, however, believes that this method leads to the same relation of simultaneity for each inertial frame *I*, to the effect that a relativization to *I* proves to be unnecessary. From the viewpoint of special relativity this is an erroneous belief. It can be explained quite easily, however, by taking into consideration that Newtonian physicists dealt with inertial frames whose relative velocities were only very small, and that they had no pressing theoretical reasons to question their familiar space-time framework. Thus, it seems perfectly plausible to say that in the situation of the cannon ball method Newtonian physicists, when using the word "simultaneous", are talking about the relation of simultaneity relative to that inertial frame in which the guns are at rest. This interpretation, given from the point of view of special relativity, allows satisfactory explanations of the beliefs and the linguistic behavior of Newtonian physicists.

Already this very simple and straightforward interpretative story indicates the implausibility of Thomas Kuhn's incommensurability claims, at least in the case of the Newton–Einstein transition. Of course, other stories of this kind, dealing with other situations, e.g. situations where light signals or transported clocks are used to determine simultaneity, can be told as well[29] and it is relatively easy to tell all of them in such a way that they do not involve any change of the reference of the term "simultaneous".

My second example concerns the notion of mass. I will deliberately leave gravity out of consideration in order not to become entangled in problems of *general* relativity. Even with this restriction, the notion of mass still proves to be complicated and interesting. In Newtonian physics "mass" is mostly understood in the sense of *inertial* mass, i.e., as a quantity measuring the inertia of a body; sometimes, however, "mass" is also understood in the sense of a measure of the amount of matter of a body. Special relativity theory, too, deals with inertial mass. In contrast to Newtonian beliefs, however, inertial mass is now regarded as dependent on the speed of a body and in particular as dependent on a frame of reference. Furthermore, in special relativity there is the so-called *proper* mass of a body, which is identical with the inertial mass of the body relative to its rest frame, and there also may be again, at least implicitly, the idea of mass as a measure of the amount of matter. The question then arises of how, from the point of view of special relativity theory, we should interpret the term "mass" as it is used by Newtonian physicists.

This question evokes quite different reactions. Physicists mostly show their typical lack of interest in semantical-historical questions of this kind ("like the question as to how many angels may dance on the point of a needle"). Many tend to regard it as unanswerable since the Newtonian notion of mass appears not to be well-defined to them. Thomas Kuhn regards it as unanswerable, too, but for another reason: He thinks that the Newtonian term "mass" refers to something which he calls "Newton-

ian mass" and which he considers incommensurable to everything special
relativity may offer.[30] Hartry Field has disputed that and taken the view
that there is no incommensurability but rather referential indeterminacy:
The Newtonian term "mass" refers, as it were, half to inertial and half to
proper mass.[31] John Earman, however, in a reply to Field, has argued
that there is neither incommensurability nor indeterminacy, but that the
interpretation as proper mass is the adequate one.[32]

A further possibility, as suggested by Philip Kitcher, though with respect
to other scientific theories,[33] would consist in distinguishing between different types of contexts (as already indicated in my foregoing example
concerning the notion of simultaneity) and in telling, for each context
type, a separate interpretative story. I have investigated this possibility in
my previous writings (especially in my 1989) and it seemed to me that
one can always tell plausible stories which bring about unique interpretations from the point of view of the Einsteinian theory, even though there
might be different ones for different sorts of contexts. Consequently, there
is constancy of reference, although relativized to context types.

To give an example, let us consider the context of collision processes.
In this context one can use a formal analysis of the Newton–Einstein
transition devised by Jürgen Ehlers and going back to ideas of Langevin,
Penrose and Rindler.[34] Its central characteristic is a formal framework,
making use of the language of differential geometry on a four-dimensional
manifold, in which both the Newtonian and the relativistic theory of
collision are expressed by essentially the same axioms, with only a single
difference: In the axioms appears a parameter λ such that, if $\lambda = 0$, one
gets the Newtonian theory, and if $\lambda = c^{-2}$, one gets the relativistic theory
(where c, as usual, is the speed of light in vacuo, or, more generally, the
fundamental speed which is invariant under Lorentz transformations). In
this framework the notorious approximations which, in certain domains
of application, make the two theories experimentally indistinguishable,
can be expressed in a formally precise and satisfying way.

With respect to the notion of mass Ehlers proceeds as follows. In the
relativistic case, he assigns to any particle p taking part in a certain
collision three different kinds of mass. First the *proper* mass $m(p)$ (which
is proportional to the energy of p relative to the rest frame of p). Second,
he allows that any particle p taking part in a collision is composed of other
particles and that these compositions may be changed by the collision. If
these other particles are 'elementary', we simply add their proper masses
and thereby get something like a measure of the amount of matter of p.
It is denoted by $M(p)$ and called *conserved* mass of p since, with regard
to all the particles taking part in a collision, the sum of all the $M(p)$
before the collision is the same as the sum after the collision.[35] The proper
mass $m(p)$, then, is the sum of the conserved mass $M(p)$ and a remainder
which represents the binding energy, needed to hold the elementary particles together, and possibly other energy contributions as well. Both $m(p)$

and $M(p)$ are independent of inertial frames. Third, there is, relative to any inertial frame I, the *inertial* mass $m_I(p)$. $m_I(p)$ is connected with $m(p)$ by the simple equation $m_I(p) = \gamma(v)m(p)$, where v is the speed of the particle p relative to I and $\gamma(v)$ the Lorentz factor depending on v.

Ehlers' analysis then shows that if we interpret the Newtonian term "mass" as *conserved* mass, within the usual domain of application of the Newtonian theory – i.e., the domain of application where all particles have small speeds and small internal energies – we can find for any model of the special relativistic theory a model of the Newtonian theory which is experimentally indistinguishable from it – and vice versa. This means, in particular, that in the case of those collision processes with which the Newtonian physicists were mostly concerned, the laws of the conservation of momentum and energy, *as formulated by the Newtonian physicists themselves*, are also true – within the margins of experimental error – from the standpoint of special relativity, *if* one interprets the Newtonian term "mass" as conserved mass. Furthermore, conserved mass itself (as its name suggests) satisfies a conservation law, and, finally, conserved mass is independent of inertial frames. Thus, it has quite a few of the properties Newtonians were presupposing with respect to mass. Consequently, it appears to be the ideal candidate for an interpretation of the Newtonian term "mass" from the standpoint of special relativity. This is the interpretation which may be proposed in the context of collision processes.

It might be objected that "conserved mass" is a dubious notion, or at least a notion with an artificially restricted domain of application (see note 35), and that it should not be used at all. It seems to me, however, that, even if it may be dubious in a purely special relativistic context, its use may nevertheless be allowed in *interpreting* the *Newtonian* term "mass" from the special relativistic standpoint. After all, it should be expected that such interpretations must involve certain artificialities. Furthermore, this interpretation in a sense complies with the Kuhnian demand – which certainly has its justification – that the Newtonian notions should be understood with their specific Newtonian characteristics. In contrast to what Kuhn thinks, however, we now see how it is possible to save this demand to a considerable extent *without* leaving special relativity. With "conserved mass" we have found a special relativistic notion – 'artificial' as it may be – which, in the context of collision processes, is able to reflect the Newtonian usage of "mass" in a rather satisfying way.

3. DEFEATIST VIEWS OF REFERENCE

Unfortunately, the importance of the foregoing result – and of interpretative stories concerning reference in general – may be disputed. The question of what Newtonians, considered from the standpoint of special relativity theory, referred to by their term "mass" can easily be dismissed.

Perhaps the interpretative stories which one is tempted to tell here are, after all, nothing but stories. Everybody tells his own pet story – but from a sober point of view, which is directed at hard matters of fact, real decisions between them may be impossible. An instrumentalist, in particular, would argue in this way. From an instrumentalist point of view there is a term "mass" – or a symbol "m" – which is used in the Newtonian and the relativistic theory. Both theories have their observable consequences, and measured against these consequences the relativistic theory turns out to be prelerable. Why, then, ask whether the term "mass" – or the symbol "m" – in both theories refers to the same quantity? Or, more generally, why ask how the reference of this term or symbol, as used by Newtonians, may be seen from the standpoint of relativity theory? Is it not quite unimportant which answers we give here?

So an instrumentalist may simply dispute the *presupposition* which I shared until now with Kuhn (and also with Field, Earman and Kitcher). It is the presupposition that questions about constancy of reference are important. If we came to consider these questions unimportant then we would be led to further destruction of the general notion of cognitive meaning. After the repudiation of intensions, now the notion of reference, i.e., of extension, would be strongly deflated as well.

At this point the suspicion may arise that with a really thoroughgoing deflation of the notion of reference an instrumentalist might get caught in contradictions, since – it may seem – this notion should be necessary at least with respect to the observations or observable entities he wants to talk about. To understand this issue, let us consider, for example, the instrumentalist Quine. Of course, Quine is not an instrumentalist throughout. He is an instrumentalist only when doing epistemology. As an ontologist, however, if he is only concerned with the constitution of the world and not with our knowledge of this constitution, he is a kind of realist in taking our best scientific theories, in all their ontological commitment, for granted (at 'face value', as he often says) and in his corresponding belief "in external things – people, nerve endings, sticks, stones" and also "in atoms and electrons and in classes".[36] Our best theories, however, do not say anything about reference and Quine, therefore, when doing ontology, says nothing about it either, at least nothing interesting and substantive. Quine as an ontologist simply adopts some sort of disquotational view of reference. The word "rabbit" refers to rabbits; and no more can be said: "At home [that is, with respect to the point of view of the ontologist who does not worry about such things as the indeterminacy of translation and reference] reference is captured by trivial paradigms in Tarski's style: 'Caesar' designates Caesar and 'rabbit' denotes rabbits, whatever *they* are. Such is face value, and in my naturalism I ask no better. See *Theories and Things*, top half of page 20".[37] This certainly is the most extreme form of deflating the notion of reference; but from Quine as an ontologist we should not expect more.

Quine's epistemological position, however, is more interesting and as an epistemologist Quine in fact appears to be a radical instrumentalist. The epistemologist Quine thinks that, ultimately, our scientific theories are nothing but instruments that help us foresee sensory perceptions and that, in particular, all our linguistic tools for reference, and even reference itself, only serve this single purpose.[38] According to this view scientific *progress*, then, consists in nothing but the fact that our theories become better instruments for empirical prediction. In other words: Scientific development is nothing more than a process of adaptation to observational data. Constancy of the reference of theoretical terms is not demanded and not necessary.

At this point the following difficulty, or even contradiction, seems to come up: Isn't it necessary, when talking about an adaptation to observational data, to presuppose that the reference of those terms that refer to the data themselves must be constant? Otherwise we would simply have lost the standard by which scientific progress might be measured. However, the observational data may be scientifically analysed and these analyses will involve further and deeper theoretical vocabulary. How could it be possible, then, to demand referential constancy in the case of observation terms and not to demand it and consider it necessary in the case of theoretical terms?

This is a difficulty indeed, but Quine does not fall prey to it since for him it is not necessary to demand constancy of the reference of observation terms. According to Quine, the sensory perceptions which our theories help us to foresee are captured by observation *sentences*, and the connection of these sentences with sense experience is constituted not by the *terms* they are compounded of but by our dispositions to assent to them as wholes.[39] Given an observation sentence and given the corresponding dispositions, certain sensory perceptions will cause us to assent to the observation sentence as a whole. Quine calls this the *holophrastic contact* of observation sentences with sense experience.[40] This contact has nothing to do with reference. Consequently, nothing seems to prevent us from a far-reaching deflation of the notion of reference.

Such a deflation is to be found in the work of Quine and also, following Quine, in the work of Donald Davidson. Davidson sometimes expresses himself drastically. In discussion I have heard him say: "There is no such concept, and no such thing, as reference".[41] This is Davidson's way of stating what Quine calls the *inscrutability* or *indeterminacy* of reference, a thesis which Davidson emphatically shares with Quine.[42] According to Quine and Davidson, "reference" is a theoretical term "whose function is exhausted in stating the truth conditions for sentences".[43] The truth or falsity of a sentence depends on the reference of the words it is compounded of, and Quine and Davidson regard this contribution to truth and falsity as the sole significant function of reference. This immediately, and almost trivially, implies that reference is extremely underdetermined:

Given any extensional interpretation of a language which is consistent with the truth values of all the sentences of the language, *any* bijection defined on the domain of this interpetation will, in a quite natural and straightforward way, lead to another interpretation which is consistent with the given truth values as well. (Such a bijection is called a *proxy function*.)

In the light of this insight, what appears to be essential is only the structure which our language and theory impose upon a domain of objects, while the objects themselves can be chosen arbitrarily. The objects only serve as hooks where the predicates of our language are hung up. The individuality of these hooks is unimportant. What is important is the predicates and their interplay. Quine expresses it thus: "Structure is what matters to a theory, and not the choice of its objects".[44]

What is the role of causality in this approach? Isn't it very plausible to think – as the adherents of the *causal theory of reference* (in its ambitious form) do – that the reference of a term is determined by causal relations between the speaker's use of the term and certain objects, namely those objects which are at the beginning of a causal chain which, when it reaches the speaker, leads to his use of the term? Davidson and Quine say "no". Causality is in fact very important in their approach, but they do not adopt a causal theory of reference but a *causal theory of truth conditions* (or, maybe, assertability conditions) for sentences.[45] Objects and processes in the world cause us to assent to or to dissent from *sentences*; the reference of the *terms* these sentences are compounded of is seen as unaffected by causal relations. To put it differently: If somebody assents to the sentence "There is a rabbit" because of his *contact* with a rabbit, nothing can be inferred about the *reference* of his word "rabbit". Our contact with reality concerns sentences and their truth conditions, and it must not be confused with what our words refer to.[46]

Not only causal relations but also the notorious baptismal ceremonies relied on by the causal theorist will not help here.[47] *Pointing*, say, to a rabbit while uttering the sentence "This is a rabbit", or "This is Harvey", at best can correlate the sentence as a whole to a certain situation (or to certain sensory stimulations, as Quine would say), but not the word "rabbit", or "Harvey", to the object (namely, a rabbit). A belief to the contrary very often hides an unconscious belief in magic. As Wittgenstein remarked in his *Philosophical Investigations*, the act of naming may all too easily be seen as a sort of occult process: "Naming appears as a *queer* connexion of a word with an object. – And you really get such a queer connexion when the philosopher tries to bring out *the* relation between name and thing by staring at an object in front of him and repeating a name or even the word "this" innumerable times. For philosophical problems arise when language *goes on holiday*. And *here* we may indeed fancy naming to be some remarkable act of mind, as it were a baptism of an object."[48] Although Quine (not, however, Davidson!) behaves somewhat mysteriously when talking about the 'inscrutability' of reference, his true intention, like that of Davidson, actually lies in avoiding anything occult or 'queer'. They avoid it by taking a point of view that deprives the notion of reference of most of its (alleged) substance.

But isn't this point of view now very strange itself (even if not occult)? To make it appear a bit less strange, one may express it as follows: It shows us that a substantive notion of reference is *not needed*; that the really substantive roles or functions may be taken over by other notions. – But even a formulation like this may not alter our feeling that the Quinean–Davidsonian viewpoint is even more strange and more counter-intuitive than, say, the Kuhnian one. After all, for Kuhn reference is still important insofar as he considers it worthwhile to claim a *change* of reference during scientific revolutions. Quine and Davidson, instead, let this whole debate about constancy or non-constancy of reference appear to be a debate about nothing.

4. A ROLE FOR THE NOTION OF REFERENCE IN THE PHILOSOPHY OF SCIENCE?

Is what Quine and Davidson say true? I do not know. I consider it an open question. To get a better feeling of the difficulties, let us consider again, from the point of view of relativity theory, the Newtonian physicist. Let us imagine that he measures the masses of a certain set of particles by involving the particles in sufficiently many collisions, by measuring all the velocities of the particles and by calculating the masses of the particles by means of the momentum conservation law. Imagine, furthermore, that he comments on what he is doing by a description like the one just given. Then, according to the interpretation which I proposed in Section 2, we should say that with the word "mass" he referred to conserved mass. Quine and Davidson, however, may ask: What is the role, the function, of this talk about a 'reference to something'? What is its substance?

In the literature it has been claimed, or at least insinuated, that this substance may show itself if we are going to say something about the speaker's successes or failures in getting about the world.[49] If, for example, the Newtonian physicist makes use of the masses determined in the way just described, in order to calculate, again with the help of the momentum conservation law, some velocities of the particles in further collisions, then the relativists can say under which circumstances these calculations will be empirically confirmed and under which circumstances they will be empiricially disconfirmed. In the latter case, even the extent of disconfirmation, i.e., the numerical difference between the calculations and the actual magnitudes, can be specified. Relativity theory thereby not only foresees the successes and failures of the Newtonian approach, but also *explains* them.

Let us consider these explanations in more detail. We know that the

Newtonian physicists, when dealing with particles with small speeds and small internal energies, in general were successful; and we can explain these successes quite simply by observing that, under the conditions stated, the Newtonian formulas turn into the relativistic ones. No wonder, then, that the Newtonian physicists were successful. Explanations of this sort, however, need not bring *reference* into play. It is enough to say how the Newtonian physicists *applied* their formulas, how they used them as *instruments* – and more does not seem to be necessary.

Is it of any help here to consider not only the *actions* of Newtonian physicists but also their propositional attitudes, for example their *beliefs* and *desires*? The Newtonian physicists desire empirically correct results, and they believe they get these results, if they assume conservation laws for mass, momentum and energy, and if they assume, furthermore, that mass is independent of inertial frames. If we interpret their term "mass" as conserved mass, these beliefs, as we have seen, in many cases turn out to be true also from the standpoint of special relativity theory. Doesn't *this* fact speak in favour of the hypothesis that with their term "mass" Newtonians refer to conserved mass, and not to anything else?

Alas, this question does not have a straightforward answer, since we now find ourselves in the opaque debate about the status of propositional attitudes. Quine, notoriously, simply dismisses the idioms of propositional attitude – and intensional notions in general – as nonscientific.[50] And Davidson understands these idioms in such a way that the indeterminacy of reference is not affected. Quine's and Davidson's positions still appear strange to me, and it is not so easy to repress the feeling that the notion of reference *should* play a more substantial role – at least in philosophy of science – than allowed by them.[51] But at the same time it seems to me that we should reckon with the fact that philosophy might really have to teach us something important at this point, and, maybe, for a philosopher this fact should not be *so* alarming.

If Quine and Davidson were right, what would be the consequences? At first sight it might seem that the interpretative stories which Kitcher is so fond of, and which I was fond of, too, would then be totally invalidated. But this cannot be true. There is too much scientific material which flows into these stories to leave them without substance. We have to understand, instead, what their substance really is. If reference is deflated, what do these stories tell us?

In the light of the Davidsonian view, there is an obvious answer to this question: These stories tell us something, not about reference, but about the truth conditions (or assertability conditions) of the statements uttered by the former scientists judged from the viewpoint of the later scientists. And the telling of these stories should now be seen in analogy to the Davidsonian project of interpretation, with the important difference, however, that in our scientific context the Davidsonian requirement that an interpretation has to be *radical* – which means, in first approximation, that the interpreter's data about the speaker to be interpreted must not contain anything semantical – may be considerably weakened since we are now not interested in semantics ('theory of meaning') as such but rather in specific semantical questions that are relevant to a proper understanding of science and its development. Our task would be, then, to give an account of the specific aims and properties of this weakened conception of interpretation within philosophy of science.

Of course, I cannot even begin this task here. It seems to me, however, that it may reopen substantive questions about reference that have faded out in the purely semantical context of Davidson's endeavour. For in our scientific context we may simply *presuppose* the fact that former and later theoreticians – think of Newtonian and Einsteinian physicists share a considerable part of their vocabulary, reference included.[52] We can ask, then, from the point of view of the later theory and *relative* to the reference of these shared terms, what the former theorists referred to by the other, problematic terms. In this setting the 'inscrutability of reference' can no longer be deduced since we now treat the reference of the shared terms as fixed, such that the Quinean–Davidsonian reinterpretations by means of proxy functions, which in principle involve *all* the terms of our language, may be blocked.

This project of *relative* interpretation may be seen in analogy to empiricist reduction programs, where our 'shared terms' correspond to the empiricist's 'observational terms' and our 'problematic terms' to the empiricist's 'theoretical terms'. It is a program, however, which is considerably more liberal, since it does not make use of an epistemologically ambitious observational-theoretical distinction and since it does not demand that the problematic terms of the former theory are defined – or 'reduced' in some precise logical sense – by means of other terms. It only demands that we can give referential interpretations by telling plausible interpretative stories which may presuppose that the reference of the shared terms remains unchanged. Perhaps we will, then, discover new sides to the notion of reference showing more substance than conceded by the purely semantical approach of Quine and Davidson.[53]

NOTES

[1] See the texts collected in Dürr (1986).

[2] And by the members of the Berlin Society of Empirical Philosophy as well. There is a tendency (which may be historically justified) to reserve the label "Logical Empiricists" for the latter and to call the members of the Vienna Circle "Logical Positivists". I will ignore this distinction and use the less ambitious label "Logical Empiricism" for the kind of empiricism common to both groups.

[3] In what follows, the term "semantical" will always be used in this sense which should not be confused with what some people – rather misleadingly, to my mind – call the 'semantic

conception of theories', according to which theories are identified with models or classes of models (see, e.g., Suppe 1988 or van Fraassen 1989, Chapter 9). I will not deal with this conception.

[4] See Reichenbach (1951) for a relatively sophisticated version of this theory.

[5] Quine (1986a, pp. 155f).

[6] Quine thinks that his 'observation sentences' are exceptions and he offers us as their meanings what he calls "stimulus meaning"; see Quine (1960, Sections 8–10), and Quine (1992, Sections 1f and 15f). "Stimulus meaning", however, is a rather strange notion of meaning as can be seen, e.g., by the fact that stimulus meanings have to be relativized to single persons and that they are not invariant under normal processes of belief fixation and change of belief. For the latter critique – which may be directed at verificationism in general – and also for meaning holism in general, see Putnam (1986, 1987). In what follows I will ignore the issue of stimulus meanings.

[7] See Quine (1981, p. 27), and Quine (1992, Section 4).

[8] See Quine (1981, pp. 24–30).

[9] Hempel (1988).

[10] In *Pursuit of Truth* Quine concedes that his conception of empirical content is unrealistic. He now calls theories possessing empirical content in his sense "testable theories", and he writes: "much solid experimental science fails of testability in the defined sense. This can happen [. . .] because of vague and uncalibrated probabilities in the backlog of theory. No doubt it happens also in more complex ways, not clearly understood. I have no definition of empirical content to offer for such theories" (Quine 1992, p. 95). Nevertheless, he still thinks that his idealized picture "brings out the essential relation between scientific theory and its evidence" (Quine 1993, p. 111).

[11] *Philosophical Investigations*, Section 118; in G. E. M. Anscombe's English version of this section in Wittgenstein 1953, Wittgenstein's word "Luftgebäude" is translated as "house of cards" .

[12] The term "theory-laden" stems from Hanson (1958, p. 19). As pointed out by Michael Friedman (1992), the theory-ladenness of observation sentences has been explicitly stated, from the very beginning of their movement, also by the Logical Empiricists. This insight, however (as also noted by Friedman, p. 92), is simply belied by their verificationism.

[13] See, however, (the early) Feyerabend's attempt, in Feyerabend (1958), to save this dichotomy, at least to a certain extent, by proposing a pragmatic notion of observation sentence which very much resembles the Quinean one. See also Feyerabend's later assessment of this notion in his 'Nachtrag 1977' in Feyerabend (1978, pp. 24–33).

[14] All quotations from Kuhn (1970, p. 102).

[15] Kuhn (1970, p. 102).

[16] I gather this mainly from the writings of Quine (who has more to offer than a mere *reductio ad absurdum* of verificationism), Davidson and Putnam. See also Schiffer (1987) for a very pessimistic outlook on the notion of meaning in the sense of intension.

[17] We should get along without *using* it. Whether one should also plead for abandoning further *investigations* of that notion is another question on which I will not take a position.

[18] "'meaning' is not the rubric under which incommensurability is best discussed", at least "in the present state of the theory of meaning" (Kuhn 1983, p. 671).

[19] See Kuhn (1983).

[20] "I was wrong to speak of translation" (Kuhn 1993, p. 324).

[21] See Kuhn (1993).

[22] I borrow this formulation from Parsons (1982, p. 497). This notion of reference may sound a bit technical, but to my mind it only expresses the more or less obvious core of what we colloquially mean by "reference". In what follows, everything which I will *presuppose* about reference will be more or less obvious from the point of view of our colloquial notion. The interesting point then is that, in the end, philosophical reflection may lead us to *consequences* which are anything but obvious.

[23] See Quine (1961).

[24] See again Parsons (1982, pp. 497f).

[25] This is Putnam's example (1975, p. 197, and 1988, pp. 12f).

[26] See Kuhn (1970, Chapter X, and his comments thereupon in Kuhn 1993).

[27] More detailed versions of these and other interpretative stones are to be found in my 1989 (Section 7.4), and I have been inspired to tell such stories by Kitcher (1978, 1982). Kitcher sticks to this approach to this day; see his 1993 (pp. 76–80, 95–105).

[28] It may be objected that no Newtonian physicist actually used, or even thought of, such a method since he did not consider it necessary to do so. To a certain extent, this may be true. But any Newtonian physicist, if presented with such a method, would have accepted it as a legitimate method of determining simultaneity; and that is enough for my present purpose. Furthermore, there are, at least in principle, genuinely *Newtonian* methods of determining simultaneity, for example methods making use of gravitational interaction which was thought to be instantaneous and therefore constitutive of simultaneity, a point which is in fact explicitly recognized in Kant's view of Newtonian physics. (Michael Friedman reminded me of this fact.) Of course, from the relativistic point of view these methods involving gravitation must be considered inappropriate. Nevertheless, they show that Newtonians have not been as dogmatic or unreflective as they are sometimes described.

[29] As I have done in my 1989 (Section 7.4).

[30] See, as already mentioned, Kuhn (1970, p. 102).

[31] See Field (1973).

[32] See Earman (1977).

[33] See Kitcher (1978, 1982).

[34] See Ehlers (1986) and the literature cited therein.

[35] This seems only to be true, however, if in the collisions there are no transmutations of the elementary particles themselves. In what follows I will always presuppose that this is the case. Of course, this presupposition very much restricts the applicability of Ehlers' analysis. (I am grateful to Heinz–Jürgen Schmidt for calling my attention to this point.) It should be noted, by the way, that "conserved mass" is a basic notion in Ehlers' axioms and that the other notions of mass – "proper mass" and "inertial mass" – are definable. This fact, however, is of no importance to the use I will make of these axioms.

[36] Quine (1981, p. 21).

[37] Quine (1986b, p. 367).

[38] Compare Quine (1981, pp. 1, 20). Strictly speaking, Quine does not talk about sensory perceptions but about 'triggerings of our sensory receptors'; but, as we will see shortly, in the present context this difference may be ignored.

[39] Accordingly, the instrumentalist Quine does not need a distinction between observation and theoretical *terms* and he can permit himself, with his notion of an observations *sentence*, to connect 'observation' with something as theoretical as the triggerings of our sensory receptors (see the foregoing note).

[40] See Quine/Ullian (1978, Chapter III), and Quine (1986c, p. 428).

[41] In his published work we come across a paper entitled 'Reality Without Reference', in which he says: "We don't need the concept of reference; neither do we need reference itself" (Davidson, 1984, p. 224).

[42] See Quine (1981, pp. 1–23), Quine (1992, Section 20), and Davidson (1993, p. 54).

[43] Davidson (1984, p. 223).

[44] Quine (1981, p. 20). Of course, results of this sort are not only to be found in the work of Quine and Davidson but also (at least as intermediary results) in the work of, for example, Frege, Schlick, Russell, Carnap, Weyl and Putnam. As regards Frege, see Dummett (1991, pp. 211–222); as for Schlick, Russell, Carnap and Putnam, see Demopoulos/Friedman (1985); Hermann Weyl expresses himself as follows: "A science can only determine its domain of investigation up to an isomorphic mapping. In particular it remains quite indifferent as to the 'essence' of its objects. [. . .] The idea of isomorphism demarcates the self-evident insurmountable boundary of cognition" (Weyl, 1949, pp. 25f). It would lead me too far away to go into the views of all these writers here, though it might be an instructive task

to analyze and compare their specific indeterminacy-producing trains of thought. They can be very different from the Quinean–Davidsonian one.

[45] See the illuminating discussion in Fodor/Lepore (1992, pp. 155–161). (Of course, Fodor and Lepore do not agree with Quine and Davidson.)

[46] See Davidson (1993, p. 55). Compare with that what Kitcher says in his (1993, p. 76): "Imagine that a speaker produces a token of a term, "the morning star", say, and thereby refers to an object. Somehow a connection is made between the noises that the speaker produces and a part of nature". Quine and Davidson, of course, would reply that the token of the term "the morning star" will normally be part of a token of a sentence – "There is the morning star", say – and that it is this *sentence* token which is 'connected to a part of nature'. The question as to what the *term* token refers to has to be answered only afterwards.

[47] Kitcher takes over this dubious part of the causal theory of reference in an astonishingly uncritical manner. According to him "a first user [introduces a] term by uttering it in the presence of an object with a present intention to give the name to that object" (Kitcher 1993, p. 77). This, however, certainly will not do as a theory of reference since the 'intention to give the name to that object' already presupposes reference to that object; see Putnam (1981, pp. 1f, 42f).

[48] Wittgenstein (1953, Section 38).

[49] See Papineau (1990, especially note 10 on p. 29). Papineau's main concern is a substantive notion of truth, but he seems to think that this involves a substantive notion of reference, too. See also Field (1992, pp. 32ff.) for some suggestions concerning the importance of explanations of success and failure.

[50] See, e.g., Quine (1960, Chapter VI). On page 218, for example, he writes: "In general the underlying methodology of the idioms of propositional attitude contrasts strikingly with the spirit of objective science at its most representative."

[51] Hilary Putnam seems to think that his 'internal realism' may offer a way out here; see Putnam (1981, especially Chapters 2, 3 and 5). But this is not at all clear to me. In Chapter 2, he devises very good defences of the Quinean thesis of the indeterminacy of reference, and if in Chapter 3 he explicitly discusses reference within internal realism, he seems to be driven to nothing more than a disquotational view of reference. Whether his internal realism makes room for more than that – as insinuated in Chapter 5 – remains very unclear. If I understand him correctly, Putnam is searching for a substantive intermediate position between a disquotational view of reference, on the one hand, and referential indeterminacy, on the other, but I cannot see – also in his writings after 1981 – that he has succeeded in really approaching such a position.

[52] This is, albeit reluctantly, also conceded by Kuhn: "Most of the terms common to the two theories function the same way in both; their meanings [reference included; F.M.], whatever they may be, are preserved; their translation is simply homophonic" (Kuhn 1983, p. 670).

[53] I am grateful to the members of the research group *Semantical Aspects of Spacetime Theories* for their valuable comments on an earlier version of this paper and to Emily Carson and Michael Hallett who corrected and improved the English of the final version.

REFERENCES

Davidson, D.: 1984, *Truth and Interpretation*, Clarendon Press, Oxford.

Davidson, D.: 1993, 'Reply to Felix Mühlhölzer", in R. Stoecker (ed.), *Reflecting Davidson*, Walter de Gruyter, Berlin, pp. 54f.

Demopoulos, W. and M. Friedman: 1985, 'Bertrand Russell's *The Analysis of Matter*: Its Historical Context and Contemporary Interest', *Philosophy of Science* 52, 621–639.

Dürr, H.-P. (ed.): 1986, *Physik und Transzendenz*, Scherz, Bern.

Dummett, M.: 1991, *Frege: Philosophy of Mathematics*, Duckworth, London.

Earman, J.: 1977, 'Against Indeterminacy', *J. Phil.* 74, 535–538.

Ehlers, J.: 1986, 'On Limit Relations Between, and Approximative Explanations of, Physical Theories', in R. Barcan Marcus et al. (eds.), *Logic, Methodology and Philosophy of Science VII*, Elsevier, Amsterdam, pp. 386–403.

Feyerabend, P.: 1958, 'An Attempt at a Realistic Interpretation of Experience', *Proc. Arist. Soc., n.s.* **58**, 143–170.

Feyerabend, P.: 1978, *Der wissenschaftstheoretische Realismus und die Autorität der Wissenschaften (Ausgewählte Schriften, Band 1)*, Vieweg, Braunschweig.

Field, H.: 1973, 'Theory Change and the Indeterminacy of Reference', *J. Phil.* **70**, 462–481.

Field, H.: 1992, 'Critical Notice: Paul Horwich's *Truth*', *Philosophy of Science* **59**, 321–330.

Fodor, J. and E. Lepore: 1992, *Holism – A Shopper's Guide*, Blackwell, Oxford.

Friedman, M.: 1992, 'Philosophy and the Exact Sciences: Logical Positivism as a Case Study', in J. Earman (ed.), *Inference, Explanation, and Other Frustrations*, University of California Press, Berkeley, pp. 84–98.

Hanson, N. R.: 1958, *Patterns of Discovery*, Cambridge University Press, Cambridge.

Hempel, P.: 1988, 'Provisos: A Problem Concerning the Inferential Function of Scientific Theories', in A. Grünbaum and W. C. Salmon (eds.), *The Limitations of Deductivism*, University of California Press, Berkeley, pp. 19–36.

Kitcher, P.: 1978, 'Theories, Theorists and Theoretical Change', *Phil. Rev.* **87**, 519–547.

Kitcher, P.: 1982, 'Genes', *Brit. J. Phil. Sci.* **33**, 337–359.

Kitcher, P.: 1993, *The Advancement of Science*, Oxford University Press, New York.

Kuhn, T. S.: 1970, *The Structure of Scientific Revolutions*, 2nd ed., The University of Chicago Press, Chicago.

Kuhn, T. S.: 1983, 'Commensurability, Comparability, Communicability', in P. Asquith and T. Nickles (eds.), *PSA 1982*, Vol. 2, Philosophy of Science Association, East Lansing, pp. 669–688.

Kuhn, T. S.: 1993, 'Afterwords', in P. Horwich (ed.), *World Changes*, MIT Press, Cambridge, Massachusetts, pp. 311–341.

Mühlhölzer, F.: 1989, *Objektivität und Erkenntnisfortschritt – Eine Antwort auf Thomas S. Kuhn*, Habilitationsschrift, Universität München.

Papineau, D.: 1990, 'Truth and Teleology', in D. Knowles (ed.), *Explanation and its Limits*, Cambridge University Press, Cambridge, pp. 21–43.

Parsons, C.: 1982, 'Objects and Logic', *The Monist* **65**, 491–516.

Putnam, H.: 1975, *Mind, Language and Reality, Philosophical Papers Vol. 2*, Cambridge University Press, Cambridge.

Putnam, H.: 1981, *Reason, Truth and History*, Cambridge University Press, Cambridge.

Putnam, H.: 1986, 'Meaning Holism', in L. E. Hahn and P. A. Schilpp (eds.), *The Philosophy of W. V. Quine*, Open Court, La Salle, Illinois, pp. 405–426.

Putnam, H.: 1987, 'Meaning Holism and Epistemic Holism', in K. Cramer, H. F. Fulda, R.-P. Horstmann and U. Pothast (eds.), *Theorie der Subjektivität*, Suhrkamp, Frankfurt, pp. 251–277.

Putnam, H.: 1988, *Representation and Reality*, MIT Press, Cambridge, Massachusetts.

Quine, W. V.: 1960, *Word and Object*, MIT Press, Cambridge, Massachusetts.

Quine, W. V.: 1961, 'Notes on the Theory of Reference', in W. V. Quine, *From a Logical Point of View*, 2nd ed., Harper & Row, New York, pp. 130–138.

Quine, W. V.: 1981, *Theories and Things*, Harvard University Press, Cambridge, Massachusetts.

Quine, W. V.: 1986a, 'Reply to Roger F. Gibson, Jr.', in L. E. Hahn and P. A. Schilpp (eds.), *The Philosophy of W. V. Quine*, Open Court, La Salle, Illinois, pp. 155–157.

Quine, W. V.: 1986b, 'Reply to Robert Nozick', in L. E. Hahn and P. A. Schilpp (eds.), *The Philosophy of W. V. Quine*, Open Court, La Salle, Illinois, pp. 364–367.

Quine, W. V.: 1986c, 'Reply to Hilary Putnam', in L. E. Hahn and P. A. Schilpp (eds.), *The Philosophy of W. V. Quine*, Open Court, La Salle, Illinois, pp. 427–431.

Quine, W. V.: 1992, *Pursuit of Truth*, revised edition, Harvard University Press, Cambridge, Massachusetts.

Quine, W. V.: 1993, 'In Praise of Observation Sentences', *J. Phil.* **90**, 107–116.
Quine, W. V. and J. S. Ullian: 1978, *The Web of Belief*, 2nd ed., Random House, New York.
Reichenbach, H.: 1951, 'The Verifiability Theory of Meaning', *Proceedings of the American Academy of Arts and Sciences* **80**, 46–60.
Schiffer, S.: 1987, *Remnants of Meaning*, MIT Press, Cambridge, Massachusetts.
Suppe, F.: 1988, *The Semantic Conception of Theories and Scientific Realism*, University of Illinois Press, Urbana, Illinois.
van Fraassen, B. C.: 1989, *Laws and Symmetry*, Clarendon Press, Oxford.
Weyl, H.: 1949, *Philosophy of Mathematics and Natural Science*, Princeton University Press, Princeton.
Wittgenstein, L.: 1953, *Philosophical Investigations*, Blackwell, Oxford.

Technische Universität Dresden
Institut für Philosophie
D-01062 Dresden
Germany

JOHN D. NORTON

DID EINSTEIN STUMBLE? THE DEBATE
OVER GENERAL COVARIANCE

ABSTRACT. The objection that Einstein's principle of general covariance is not a relativity principle and has no physical content is reviewed. The principal escapes offered for Einstein's viewpoint are evaluated.

1. INTRODUCTION

... the general theory of relativity. The name is repellent. Relativity? I have never been able to understand what that word means in this connection. I used to think that this was my fault, some flaw in my intelligence, but it is now apparent that nobody ever understood it, probably not even Einstein himself. So let it go. What is before us is Einstein's theory of gravitation. (Synge 1966, p. 7)

The magnitude of Einstein's success with his theories of relativity brought its own peculiar problem. His success attracted legions of cranks to his work, all determined to show where Einstein had blundered and anxious to accuse him of the most fundamental of misconceptions. On first glance, you might well imagine that the sentiments quoted above were drawn from this tiresome crank literature. However you would be mistaken. These remarks were made by J. L. Synge, one of this century's most important and influential relativists. They reflect the growth of a tradition of criticism of Einstein's views on the foundations of general relativity. The tradition began with the theory's birth in the 1910s as a minority opinion. Over the decades following, it refused to die out, instead growing until it is now one of the major schools of thought, if not the majority view amongst relativists.

The deep reservations of this tradition do not apply to the theory itself. The general theory of relativity is nearly universally hailed as our best theory of space, time and gravitation and a magnificent intellectual achievement – although followers of Synge might prefer a different name for the theory. What is questioned is the account that Einstein gave of its fundamental postulates. His account has been criticized in many of its aspects. The one that has attracted the most criticism is the prominence he accorded the requirement of general covariance, which Einstein saw as the crowning achievement of his theory. Through it, Einstein proclaimed, the theory had extended the principle of relativity to accelerated motion. Einstein's critics responded that general covariance had nothing to do with a generalization of the principle of relativity. Worse, general covariance was physically vacuous, a purely mathematical property.

My purpose in this paper is to review some of the principal positions advanced in this debate.[1] I will pursue two themes: whether covariance

103

Erkenntnis **42**: 223–245, 1995.

JOHN D. NORTON

principles have physical content and whether they express a relativity
principle. First, in Sections 2 and 3, I will review the role Einstein claimed
for covariance principles in the foundations of relativity theory and the
ensuing objection, originating with Kretschmann in 1917, that the principle
of general covariance is physically vacuous. Then, in Section 4, I will
outline the stratagems that have been proposed to restore physical content
to the principle. I will conclude that they succeed only in the degree to
which they deviate from a simple reading of the original principle. In
Section 5, I will review the development of the modern view that covariance principles are not relativity principle and that relativity principles
express a symmetry of a spacetime. Finally, in Section 6, I will review
Anderson's notion of absolute object. This notion provides our best attempt to reconcile Einstein's view of the connection between covariance
and relativity principle and the modern view of relativity principles as
symmetry principles.

2. COVARIANCE IN EINSTEIN'S ACCOUNT OF THE FOUNDATIONS OF RELATIVITY THEORY

For Einstein, covariance principles were the essence of his theories of
relativity. For a theory to satisfy a principle of relativity, the equations
expressing its laws needed to have a particular formal property. They
needed to remain unchanged in form – covariant – under a group of
coordinate transformations characterizing the principle of relativity at
issue. This was the clear moral of his famous 1905 special relativity paper
(Einstein 1905). The emphasis in that paper was to discover the correct
form of the group of coordinate transformations associated with the relativity of inertial motion. These, he argued, were the Lorentz transformation equations. It then followed that Maxwell's electrodynamics satisfied
the principle of relativity of inertial motion since the basic equations of
Maxwell's theory remained unchanged in form under Lorentz transformation. Einstein (1940, p. 329) later summarized his approach:

The content of the restricted relativity theory can accordingly be summarized in one sentence:
all natural laws must be so conditioned that they are covariant with respect to Lorentz
transformations.

That the laws of a theory have the appropriate covariance is something
that must be demonstrated by calculation, often by quite arduous manipulation. The mechanical exercise of establishing the Lorentz covariance of
Maxwell's theory occupies a significant part (§§6, 9) of Einstein's 1905
paper.
  Einstein's algebraic approach to the principle of relativity was quite
different from that soon introduced by Minkowski (1908, 1909). He formulated the special theory of relativity in terms of the geometry of what we
now know as a Minkowski spacetime. Satisfaction of the principle of

relativity of inertial motion followed automatically provided one used only the natural geometric structures of the Minkowski spacetime to formulate one's theory. Sommerfeld's (1910, p. 749) capsule formulation of what it took to satisfy the principle of relativity was quite unlike Einstein's:

According to Minkowski, as is well known, one can formulate the content of the principle of relativity as: Only *spacetime vectors* may appear in physical equations . . . .

Instead of tedious calculation to verify preservation of form of equations under transformation, one could verify that a theory formulated in a Minkowski spacetime satisfied the principle of relativity by inspection.

In 1907, Einstein began the long series of investigations that would ultimately lead to his general theory of relativity. Einstein's goal was to construct a relativistically acceptable gravitation theory by extending the principle of relativity to acceleration.[2] Throughout the entire project Einstein's emphasis remained on covariance principles and the associated algebraic viewpoint. His first step was to unveil (Einstein 1907, Part V) what he hoped would be the key to the extension, the hypothesis of the complete physical equivalence of uniform acceleration in a gravitation free space and rest in a homogenous gravitational field. The significance of this hypothesis – soon to be called the "principle of equivalence" – lay in the fact that it allowed Einstein to include transformations to accelerating coordinate systems in the covariance group of his theory. That is, it took the first step in extending the covariance of his special theory.[3]

This is the gist of the principle of equivalence: In order to account for the equality of inert and gravitational mass within the theory it is necessary to admit non-linear transformations of the four coordinates. That is, the group of Lorentz transformations and hence the set of "permissible" coordinate systems has to be extended. Einstein (1950, p. 347)

The completion of the project lay in the further extension of the covariance of his gravitation theory.

Even though Minkowski's spacetime approach would provide the formal basis for the final theory, Einstein was very slow to adopt Minkowski's methods. He did not use Minkowski's spacetime methods in developing his static theory of gravitation in the 1911 and 1912 (Einstein 1911, 1912a, b, c). The spacetime approach entered his analysis only with the publication of Einstein and Grossmann (1913), the first sketch of the general theory of relativity. This paper was distinctive in using the absolute differential calculus of Ricci and Levi-Civita (1901), later known as the tensor calculus. Yet Einstein's emphasis remained on the covariance properties of the laws of his theory, rather than its intrinsic geometric properties. Here Einstein and Grossmann were actually following the approach of Ricci and Levi-Civita. The latter preferred to think of their methods as providing an abstract calculus for manipulating systems of variables; the natural application in the geometry of curved surfaces was just one of many applications and was not to be allowed to dominate the method.

With the writing of Einstein and Grossmann (1913), Einstein's quest for his general theory of relativity should have been completed, for he had virtually the entire theory in hand. Their theory differed only from the final theory in its gravitational field equations. For reasons that have been dissected extensively elsewhere (Norton 1984), Einstein and Grossmann considered generally covariant field equations but rejected them in favor of a set of equations of restricted covariance. This disastrous decision dominated the next three years of Einstein's work on gravitation as he struggled to reconcile himself with his misshapen theory. These efforts led Einstein to his ingenious "hole argument" which purported to show that generally covariant field equations would be physically uninteresting. To show this, Einstein assumed that the gravitational field equations were generally covariant and considered a matter free region of spacetime, the "hole". He then showed that general covariance allowed him to construct two solutions of the gravitational field equations g_{ik} and g'_{ik} in the *same* coordinate system such that g_{ik} and g'_{ik} agreed outside the hole but came smoothly to disagree within the hole. This, Einstein felt, was a violation of the requirement of determinism, for the fullest specification of both gravitational field and matter distribution outside the hole must fail to fix the field g_{ik} within the hole. This violation was deemed fatal by Einstein to generally covariant gravitational field equations.[4]

In November 1915, Einstein finally emerged victorious from his struggle. He had returned to the quest for generally covariant gravitational field equations and had found the equations now routinely associated with his theory. Early the following year he celebrated his achievement with the well-known review of his new theory, Einstein (1916). Its early sections laid out the motivation and physical basis of the new theory, culminating in the association of the generalized principle of relativity with the general covariance of the theory (§3):

> The general laws of nature are to be expressed by equations which hold good for all systems of co-ordinates, that is, are co-variant with respect to any substitutions whatever (generally covariant).
>
>   It is clear that a physical theory which satisfies this postulate will also be suitable for the general postulate of relativity. For the sum of *all* substitutions in any case includes those which correspond to all relative motions of three-dimensional systems of co-ordinates. (Einstein's emphasis)

The passage continued to state what John Stachel has labelled the "point-coincidence argument".

> That this requirement of general co-variance, which takes away from space and time the last remnant of physical objectivity, is a natural one, will be seen from the following reflexion. All our space-time verifications invariably amount to a determination of space-time coincidences. If, for example, events consisted merely in the motion of material points, then ultimately nothing would be observable but the meetings of two or more of these points. Moreover, the results of our measurings are nothing but verifications of such meetings of the material points of our measuring instruments with other material points, coincidences

between the hands of a clock and points on the clock dial, and observed point-events happening at the same place and the same time.

The introduction of a system of reference serves no other purpose than to facilitate the description of the totality of such coincidences. We allot to the universe four space-time variables x_1, x_2, x_3, x_4, in such a way that for every point-event there is a corresponding system of values of the variables $x_1 \ldots x_4$. To two coincident point-events there corresponds one system of values of the variables $x_1 \ldots x_4$, i.e. coincidence is characterized by the identity of the co-ordinates. If, in the place of the variables $x_1 \ldots x_4$, we introduce functions of them, x'_1, x'_2, x'_3, x'_4, as a new system of co-ordinates, so that the system of values are made to correspond to one another without ambiguity, the equality of all four co-ordinates in the new system will also serve as an expression for the space-time coincidence of the two point-events. As all our physical experience can be ultimately reduced to such coincidences, there is no immediate reason for preferring certain systems of co-ordinates to others, that is to say, we arrive at the requirement of general co-variance.

The purpose of this argument at this point in Einstein's exposition had become completely obscured until rediscovered by Stachel (1980). The point coincidence argument was Einstein's answer to the hole argument – although Einstein only explained this in private correspondence. Readers of Einstein (1916) would have to figure this out for themselves and fill in the details alone. In brief, the argument's basic assumption was that the physical circumstance described by any field g_{ik} was exhausted by the catalog of spacetime coincidences that it allowed. It turned out that, by construction, the two fields g_{ik} and g'_{ik} of the hole argument agreed on all such coincidences. Therefore any difference between them could not be physical. It was a purely mathematical effect, one we would now describe as a gauge freedom. Thus the indeterminism of the hole argument provided no physical ground for rejecting generally covariant field equations. Finally, to forgo general covariance and the use of arbitrary coordinate systems is to restrict the theory in a way that goes beyond its physical content, for that content is exhausted by the catalog of spacetime coincidences, which cannot pick between different coordinate systems.

There was an ease in the writing of other parts of Einstein's accounts of the foundations of this theory that proved deceptive to readers who took Einstein's discussion to represent a register of uncontroversial postulates or consequences of the theory. In his Section 2, for example, Einstein had urged that the inertial properties of a body must be fixed completely by the other bodies of the universe. This was a result that Einstein had so far only been able to recover in weak field approximation. Therefore its inclusion in the discussion was more a statement of what Einstein hoped to recover from his theory than a report of what he had recovered. When he later tried to derive the full effect, he ran into very serious problems. He initially sought to contain these problems in his cosmological work by augmenting his field equations with the notorious "cosmological term". He later abandoned as wrongheaded these attempts to recover what had then become known as "Mach's Principle".

3. IS GENERAL COVARIANCE PHYSICALLY VACUOUS?

A more immediate shock awaited Einstein. Erich Kretschmann (1917) had read and understood all too clearly the fragility of Einstein's account of the foundations of his theory. He began his paper with the remarks (pp. 575–576).[5]

The forms in which different authors have expressed the postulate of the Lorentz–Einstein theory of relativity – and especially the forms in which Einstein has recently expressed his postulate of general relativity – admit the following interpretation (in the case of Einstein, it is required explicitly): A system of physical laws satisfies a relativity postulate if the equations by means of which it is represented are covariant with respect to the group of spatio-temporal coordinate transformations associated with that postulate. If one accepts this interpretation and recalls that, in the final analysis, all physical observations consist in the determination of purely topological relations ("coincidences") between objects of spatio-temporal perception, from which it follows that no coordinate system is privileged by these observations, then one is forced to the following conclusion: By means of a purely mathematical reformulation of the equations representing the theory, and with, at most, mathematical complications connected with that reformulation, any physical theory can be brought into agreement with any, arbitrary relativity postulate, even the most general one, and this without modifying any of its content that can be tested by observation.

Einstein had used the point-coincidence argument to establish the requirement of general covariance. If coincidences are all that matters physically, then we ought to be able to use any coordinate system, since all coordinate systems will agree on spacetime coincidences. Kretschmann now objects that this argument works too well. If we accept the point-coincidence argument, then we ought to be able to use arbitrary coordinates in any theory, for the physical content of any theory ought to remain unchanged with the adoption of new coordinate systems. The challenge is merely a mathematical one: find a generally covariant formulation of the theory. Kretschmann later remarked (p. 579) that this task ought to be perfectly manageable for any physical theory given the power of such methods as Ricci and Levi-Civita's.

Kretschmann's original objection was conditioned on acceptance of the point-coincidence argument. That condition was soon dropped when his remarks were cited. Kretschmann's objection is now routinely recalled as the observation that general covariance is a purely mathematical property of the formulation of a theory. Any spacetime theory can be given generally covariant formulation; general covariance has no physical content. In this form, Kretschmann's objection has become one of the most cited and endorsed objections to Einstein's account of the foundations of the general theory of relativity.

4. GENERAL COVARIANCE HAS PHYSICAL CONTENT IF YOU . . .

If the caution and awkwardness of Einstein's (1918) reply is any guide Einstein must have been quite seriously troubled by Kretschmann's assault. He began by carefully stating the three principles upon which his theory was based: the (a) principle of relativity, (b) the principle of equivalence and (c) Mach's principle. That list already contained concessions to Kretschmann, for, in a footnote, Einstein confessed that he had not previously distinguished (a) and (c) and that this had caused confusion. Further, his statement of the principle of relativity had been reduced to its most circumspect core, far removed from the vivid thought experiments usually surrounding the principle:

(a)   *Principle of Relativity*: The laws of nature are only assertions of timespace coincidences; therefore they find their unique, natural expression in generally covariant equations.

The principle was now merely a synopsis of the point-coincidence argument itself.

However this cautious reorganization was an exercise in relabelling. It had still not escaped Kretschmann's objection that general covariance is physically vacuous. Einstein's options were extremely limited, for Kretschmann had come to his conclusion by using Einstein's own point-coincidence argument and that was an argument Einstein was unable to renounce. Einstein's response was the first of a series which conceded that general covariance *simpliciter* is physically vacuous after all. In this tradition it is urged that general covariance has physical content if it is supplemented by a further requirement. The problem is to decide what that further requirement should be. Einstein's choice is best known.

4.1. Add the Requirement of Simplicity

Einstein wrote

I believe Herr Kretschmann's argument to be correct, but the innovation proposed by him not to be commendable. That is, if it is correct that one can bring any empirical law into generally covariant form, then principle (a) still possesses a significant heuristic force, which has already proved itself brilliantly in the problem of gravitation and rests on the following. Of two theoretical systems compatible with experience, the one is to be preferred that is the simpler and more transparent from the standpoint of the absolute differential calculus. Let one bring Newtonian gravitational mechanics into the form of absolutely covariant equations (four-dimensional) and one will certainly be convinced that principle (a) excludes this theory, not theoretically, but practically!

That is, Einstein responds that the requirement of general covariance has physical content if we augment it with the additional constraint that generally covariant formulations of theories must be simple.

Einstein chose to illustrate his point by challenging readers to seek a generally covariant formulation of Newtonian theory, which Einstein supposed would be unworkable in practice. The choice proved a poor one, for it was discovered shortly by Cartan (1923) and Friedrichs (1927) that it was quite easy to give Newtonian theory an entirely workable generally covariant formulation using essentially the same techniques as Einstein. Nonetheless Einstein's escape became one of the most popular in later literature. Misner, Thorne and Wheeler (1973, pp. 302–303), for example, complete their discussion of generally covariant formulations of Newtonian theory with a recapitulation of Einstein's escape in characteristically colorful language:

> But another viewpoint is cogent. It constructs a powerful sieve in the form of a slightly altered and slightly more nebulous principle: "Nature likes theories that are simple when stated in coordinate-free, geometric language" . . . . According to this principle, Nature must love general relativity, and it must hate Newtonian theory. Of all theories ever conceived by physicists, general relativity has the simplest, most elegant geometric foundations . . . . By contrast, what diabolically clever physicist would ever foist on man a theory with such a complicated geometric foundation as Newtonian theory?

Einstein's escape works in a straightforward but limited sense. The requirement of simplicity in generally covariant formulation induces an ordering on empirically equivalent theories. The physical content arises in the assumption that the simpler theory in the ranking is more likely to be true. However this success is limited by two qualifications. First, we have no objective scheme for comparing the simplicity of two formulations.[6] In practice in individual cases, judgments of relative simplicity can be made with wide agreement. However that judgment rests on the intuitive sensibilities of the people evaluating the formulations and not on explicitly stated rules. This is hardly a comfortable basis for underwriting the physical content of a fundamental physical principle. Second, the requirement of simplicity can only direct us if we have empirically equivalent theories. In this regard, Einstein and Misner, Thorne and Wheeler's comparison of general relativity and Newtonian theory is somewhat misleading and inflates the significance of the requirement. The decision between these two theories is not based on the simplicity of the generally covariant formulations. Had the celebrated tests of general relativity failed and all experiments favored Newtonian theory, could we justify our current enthusiastic support for general relativity no matter how simple it may be?

There is a deeper worry. It is possible to give an entirely innocent explanation of the empirical success of the requirement of simplicity of generally covariant formulations. We know from other grounds that general relativity is our preferred theory of space, time and gravitation. Since it happens to have an especially simple generally covariant formulation, we naturally prefer theories with such simple formulation just as a rather cumbersome way of expressing our preference for general relativity. Our preference for them, then, is merely an accidental by- product of our preference for general relativity. This innocent views seems not to be that of Einstein. He sought to elevate the requirement of simplicity to fundamental metaphysics.[7] Elsewhere he made the celebrated proclamation (Einstein 1933):

Our experience hitherto justifies us in believing that nature is the realization of the simplest conceivable mathematical ideas. I am convinced that we can discover by means of purely mathematical constructions the concepts and laws connecting them with each other, which furnish the key to the understanding of natural phenomena. . . . the creative principle resides in mathematics.

Thus Einstein's view reverses the innocent explanation of the success of a simple generally covariant theory. The success of general relativity derives from the fact of its simple generally covariant formulation.

The difficulty with Einstein's proposal is that there is scant evidence to justify the move from the innocent explanation to Einstein's deeper metaphysics. Our experience does not justify what Einstein claims, that is, that the canon of mathematical simplicity provides the decisive heuristic guide in our search for physical theories. Our experience is that major changes in physical theory require new mathematical languages and only in that new mathematical language does the theory appear simple. In the earlier languages, the theory may be extraordinarily complicated or even inexpressible. An obvious example is Einstein's own general theory of relativity, which found simple expression only by reviving a mathematical method that had lain stagnant for over a decade. A second example is the advent of the modern quantum theory in the 1920s. The theory finally found its simplest general mathematical expression in the mathematics of operators on Hilbert spaces. Yet this new mathematics emerged quite painfully as a synthesis of many explorations: the matrix methods of Born, Heisenberg and Jordan, the wave mechanics of Schrödinger, Dirac's c and q numbers as well as major contributions from group theory.

The moral of experience is that our best theories find simple mathematical expression because of the special efforts of mathematicians and physicists to find simple mathematical expression of our best theories. There is no single, natural mathematical language in which to judge the simplicity of theories. Throughout the history of science, physicists have drawn on many different mathematical tools. Sometimes it is the then standard method; sometimes it is a revival of one that has languished; and sometimes it is one that is developed hand-in-hand with the theory or even after the new theory is well established. Because of the wide range of choice of mathematical method, any successful theory choice can be cast as the choice of the mathematically simplest. In so far as mathematical simplicity has heuristic value, it is entirely dependent on the choice of the right mathematical tool. How to make that choice is only apparent *after* the success of the theory. As a heuristic guide, mathematical simplicity is entirely dependent on the fortuitous choice of the right mathematical language, that is, on being lucky by guessing correctly.

Nature is not tuned into our mathematics. Our mathematics is adjusted painstakingly to fit nature as out understanding of nature deepens. Mathematics hardly seems to provide us with a fixed and elevated vantage point from which to direct the development of new physical theories. The vantage point mathematics provides is as mutable as our physics.

4.2. *Restrict the Addition of New Structures*

The difficulty with Einstein's escape from Kretschmann's objection is that it leads us towards a problematic metaphysics of simplicity. A more empirically motivated escape originated with several authors. Kretschmann had urged that one can take any spacetime theory and find generally covariant formulation for it. Fock (1959, p. xvi) considers this transition. He points out that one can always find a generally covariant formulation of a theory if one is allowed to introduce new auxiliary quantities arbitrarily. This easy achievement of a generally covariant formulation can be blocked if we insist that any new quantity must have proper physical basis and not be a purely mathematical artifice. Trautman (1964, pp. 122–123) and Wald (1984, p. 57) use the same example to illustrate this escape. If A_a is a covector field, then the equation that merely requires the vanishing of its first component

$$A_1 = 0$$

is clearly not generally covariant. We could, however, render this equation generally covariant by explicitly introducing the coordinate basis vector field u^a associated with the x^1 coordinate, rewriting the equation as

$$u^a A_a = 0$$

This transition is blocked by the requirement that any new quantity introduced must reflect some element of physical reality according to the theory. A coordinate basis vector field, however, just reflects some arbitrary choice of coordinate system.

Pauli (1921, p. 150) gives a more realistic example of this escape. In his transition to general relativity, Einstein took a Lorentz covariant formulation of special relativity and expanded the coordinate systems it used to include those associated with uniformly accelerated motion. In this expansion, new quantities appear, the coefficients g_{ik} of the metric tensor. What makes this transition acceptable is that these coefficients do have a physical meaning. Einstein's principle of equivalence enjoins us to interpret these coefficients as describing a gravitational field.

There is a difficulty with this proposal. The problem is that it is not so much a well articulated principle as a rule of thumb with a few suggestive examples. In particular, just how are we to distinguish between new quantities that properly reflect some element of reality and those that are merely mathematical artifices? Pauli and Weyl (1921, pp. 226–227) stress that the coefficients of the metric are distinguished by the fact that they are not given a priori. They are influenced and even determined by the metric field. But this cannot be the only criterion for identifying elements of reality.[8] If we give just special relativity a generally covariant formulation, then the Minkowski metric of spacetime is represented in arbitrary coordinate systems by components g_{ik}. What principle could deny that the Minkowski metric is properly an element of reality? But if we allow this, then we must also allow generally covariant formulation of special relativity. Similarly, we can give Newtonian spacetime theory generally covariant formulation if we are allowed to treat explicitly several familiar geometric structures in spacetime: the degenerate spacetime metric h^{ab}, the absolute time one form t_a and the affine structure ∇_a. Once again each of these structures seems to represent some proper element of physical reality. But if the augmented requirement of general covariance succeeds only in ruling out contrived examples but cannot rule out generally covariant formulations of both Newtonian theory and special relativity, then it is hard to see what interesting physical content lies in the principle.[9]

4.3. Require That There is Also No Formulation of Restricted Covariance

The last escape attempted to make more precise the intuition that there is something unnatural in forcing generally covariant formulations onto theories that are more familiar in formulations of restricted covariance. Bergmann sought to give content to the principle of general covariance by building this intuition directly into its definition. The principle becomes the injunction to prefer spacetime theories that cannot be given formulations of restricted covariance. He wrote (1942, p. 159)[10]

The hypothesis that the geometry of physical space is represented best by a formalism which is covariant with respect to general coordinate transformations, and that a restriction to a less general group of transformations would not simplify that formalism, is called *the principle of general covariance*.

Thus neither special relativity nor standard Newtonian theory is admissible in generally covariant formulation. Each also admits formulations of restricted covariance, the former is Lorentz covariant, the latter Galilean covariant.

   This proposal comes closer to achieving the desired criterion for deciding between theories. However the division is not perfect. First, it is not clear that general relativity is irreducibly generally covariant. As Bondi (1959, p. 108) had pointed out, Fock (1959) has been prominent in calling for a restriction to the covariance of general relativity by augmenting it with the harmonic coordinate condition. Second, the theories that this proposal will select as generally covariant seem unrelated to those satisfying a general principle of relativity. A simplified formulation of reduced covariance is available for a theory if that theory's spacetime structures admit symmetry transformations. Since the Lorentz boost is a symmetry of a Minkowski spacetime, special relativity can be written in simplified, Lorentz covariant formulation. Thus Bergmann's proposal directs us to associate a *smaller* covariance group with a theory the *larger* the group of its spacetime symmetries. We shall see below that symmetry groups of a spacetime theory are those that are now associated with the relativity principle a theory may satisfy. Therefore, under Bergmann's proposal, the larger the group associated with a theory's relativity principle the smaller the theory's covariance group. That is, the more relative, the less covariant. For example, consider some spacetime theory that posits a completely inhomogeneous but otherwise fixed background spacetime structure. Such a theory satisfies no relativity principle at all. All events, let alone states of motion, are distinct. Yet it will be generally covariant. Special relativity and versions of Newtonian theory are more relativistic in the sense that they at least satisfy a principle of relativity of inertial motion. Yet they are less covariant.

4.4. *Contrive to Add the Result You Want*

In general, I think all these escapes contrived. What has failed is Einstein's original vision. The principle of general covariance was to have formed the core of general relativity in the way that the requirement of Lorentz covariance was central to special relativity. This hope has been dashed. The escapes discussed here all appear to be attempts to force an outcome that the unsupplemented principle could not deliver. Clearly, by sufficiently ingenious supplements, one can force the revised principle to deliver any result we choose. However we delude ourselves if we think that general covariance delivered the result. It comes from the supplement. In some cases, the supplements carry a burden extraordinarily specific to general relativity, so that it becomes hardly surprising the principle points directly to Einstein's theory. For example, Weinberg's (1972, pp. 91–92) principle of general covariance has two parts, the second only resembling Einstein's original principle. His principle

... states that a physical equation holds in a general gravitational field, if two conditions are met:

1. The equation holds in the absence of gravitation; that is, it agrees with the laws of special relativity when the metric field $g_{\alpha\beta}$ equals the Minkowski tensor $\eta\alpha\beta$ and when the affine connection $\Gamma^{\alpha}_{\beta\gamma}$ vanishes.

2. The equation is generally covariant; that is, it preserves its form under a general coordinate transformation $x \to x'$.

One might well wonder if the second condition is needed at all.

5. ARE COVARIANCE PRINCIPLES RELATIVITY PRINCIPLES?

Whatever may the outcome of the attempts to restore physical content to the principle of general covariance, a second problem remains. Is the principle a relativity principle? A long tradition of criticism maintains that covariance principles are not relativity principles. Its origins lay in a problem apparent from the earliest moments of Einstein's general theory. On the level of simple observation, there was a significant gap between the relativity principles of the special and general theory. Under the relativity of inertial motion in the special theory, an observer in a closed chamber such as a railway car could do no experiment within the car to determine whether the car was in uniform motion or at rest. Were the car to accelerate, however, that motion would be entirely apparent to the occupants of the car through inertial effects. They would hardly need to carry out delicate experiments to detect even quite modest acceleration. Yet according to Einstein an extended principle of relativity was supposed to cover both uniform and accelerated motion of the car. It was supposedly equally admissible to imagine the accelerated car still at rest but temporarily under the influence of a gravitational field.[11] Whatever may be the covariance of special and general relativity, their relativity principles seemed to be quite different and that of general relativity quite suspect.

In the 1920s it was possible to dismiss this type of skepticism about the generalized principle of relativity as shallow or even willfully obstructive – especially given its association with the politically motivated anti-relativity movement. But it became harder to dismiss the tradition of criticism that sought to give deeper expression to this worry. Kretschmann (1917) was again one of the earliest voices of this tradition of criticism. The major part of his paper had been devoted to understanding what were the relativity principles of special and general relativity. His analysis was based on a geometric characterization of a relativity principle. The relativity principle of a spacetime theory of the type of relativity was fixed by the group of transformations that mapped the lightlike and timelike worldlines of spacetime back into themselves. In the case of special relativity, this criterion gave the expected answer: the group identified was the Lorentz group. Since motions connected by a Lorentz transformation were governed by a relativity principle, the analysis returned the relativity of inertial motion as advocated by Einstein. Kretschmann's analysis gave quite different results in the case of general relativity. Lightlike and timelike worldlines were, in general, mapped back into themselves by the identity map only. It followed that general relativity satisfied no relativity principle at all. It is a fully absolute theory.

Where Kretschmann's analysis was geometric in spirit, another approach was more algebraic and closer to Einstein's own methods. The basic worry was that the form invariance of laws required by general covariance was too weak to express a relativity principle. Sesmat (1937, pp. 382–383) for example pointed out that the Lorentz transformation stood in a quite different relation to the special theory than did general transformations to the general theory. Lorentz transformations left unchanged the basic quantities describing spacetime such as the components of the metric tensor. They remained the same functions of the coordinates.

The equations of general relativity, however, could only be preserved in form under general transformations since the components of the metric tensor were adjusted by the tensor transformation law. Under arbitrary coordinate transformation, the functional dependence of these components on the coordinates would change.

Fock (1957) synthesized both geometric and algebraic approaches. A relativity principle expressed the geometrical notion of a uniformity of spacetime, such as the lack of privileged points, directions and states of motion. In a spacetime theory of the type of special and general relativity, the group associated with such uniformity is not the group under which the metric tensor transforms tensorially. It is the group under which the components of the metric tensor remain the *same functions* of the coordinates. That is, if a transformation takes coordinates x_σ to x'_σ and the components of the metric tensor $g_{\mu\nu}$ to $g'_{\mu\nu}$, then the transformation is associated with a relativity principle if it satisfies

$$g'_{\mu\nu}(x'_\sigma) = g_{\mu\nu}(x_\sigma)$$

where the equality is read as holding for equal numerical values of x_σ and x'_σ. This is the algebraic condition that expresses the uniformity of a spacetime. It is satisfied by the Lorentz transformation in special relativity but not by the general transformations of general relativity.

Fock's analysis is essentially the one now standard in both philosophical and physical literatures, although it is now expressed in the modern intrinsic geometrical language.[12] In the case of a spacetime theory which employs a semi-Riemannian spacetime, we represent a spacetime by the pair $\langle M, g_{ab} \rangle$ where M is a four-dimensional manifold and g_{ab} a Lorentz signature metric. Fock's condition picks out a group of transformations that correspond with the symmetry group of $\langle M, g_{ab} \rangle$.[13] That symmetry group is the group of all diffeomorphisms $\{h\}$ which map the metric tensor back onto itself according to

$$g_{ab} = h^* g_{ab}$$

One quickly sees that this group is the one that is naturally associated with the relativity principle. Consider, for example, two frames of reference, each represented by a congruence of timelike worldlines, in some spacetime $\langle M, g_{ab} \rangle$. If the two frames map into each other under members of the symmetry group, then any relation between the frame and background spacetime will be preserved under the map. Thus any experiment done in each frame concerning the motion of the frame in spacetime must yield the same result. Informally, spacetime looks exactly the same from each frame. If the two frames of reference are associated with railroad cars in inertial motion in a Minkowski spacetime, then any experiment concerning the motion of one car must yield an identical result if carried out in the other. But if the frames of the two cars are not related by a symmetry transformation, then the experiments will yield different results.

This is the case in which one moves inertially and the other accelerates. The experiments reveal inertial effects which distinguish the frames.

What it is for a theory to satisfy a relativity principle is understood most simply by the geometric approach to spacetime theories initiated by Minkowski, rather than the algebraic approach favored by Einstein in which one concentrates on formal properties of the equations defining the theory. If the spacetime admits a symmetry group then the theory satisfies a relativity principle for the transformations of that group.

6. ABSOLUTE AND DYNAMICAL OBJECTS

The association of relativity principles with the symmetry group of a spacetime theory seems to leave little hope for recovering a generalized principle of relativity within Einstein's general theory whatever may be its covariance. However an avenue for doing this remains. The spacetimes of special relativity are the Minkowski spacetimes $\langle M, \eta_{ab} \rangle$ where η_{ab} is a Minkowski metric. If there are matter fields present in spacetime, their presence is encoded by adding further members to this tuple. For simplicity, we can represent these further fields by their stress-energy tensor T_{ab} A theory with models $\langle M, \eta_{ab}, T_{ab} \rangle$ satisfies a relativity principle associated with the Lorentz group exactly because a Lorentz transformation is a symmetry of the pair $\langle M, \eta_{ab} \rangle$ This is the crucial point. The Lorentz transformation need not be a symmetry of the matter fields represented by T_{ab}. More generally, we divide the model $\langle M, \eta_{ab}, T_{ab} \rangle$ into two parts: $\langle M, \eta_{ab} \rangle$ which represents the background spacetime and T_{ab} which represents the matter contained in the spacetime. Figuratively, we might write this as

$$\langle M, \eta_{ab} | T_{ab} \rangle$$

where "|" represents the cut between the spacetime container and the matter contained. The group associated with the relativity principle of the theory is the symmetry group of everything to the left of this cut |.

In general relativity, the corresponding models are $\langle M, g_{ab}, T_{ab} \rangle$ It is natural to place the cut as $\langle M, g_{ab} | T_{ab} \rangle$. But that is disastrous for Einstein's approach, for the symmetry group of $\langle M, g_{ab} \rangle$ in general relativity is, in general, the identity group. The situation would be quite different if a reason could be found for relocating the cut as

$$\langle M | g_{ab}, T_{ab} \rangle$$

Then the background spacetime would merely be the bare manifold itself. If we set aside global topological issues and consider just R^4 neighborhoods of M, then the symmetry transformations are just the of C^∞ diffeomorphisms, in effect, the arbitrary transformations Einstein associated with general covariance. If the cut were moved in this way for general relativity but not special relativity, then the modern association of symmetry groups with relativity principles would pick out the Lorentz group in special relativity and the general group in general relativity.

One might seek to justify the different placing of the cut by observing that the metric tensor g_{ab} of general relativity now represents the gravitational field. Therefore it carries mass-energy and deserves to be considered as belonging on the right side of the cut as a matter field within spacetime. As it turns out there is a related way of justifying the moving of the this cut that draws directly on Einstein's own pronouncements concerning the fundamental difference between special and general relativity. Einstein insisted that a major achievement of the transition from special to general relativity was the elimination of a causal defect in the structure of special relativity. As he explained in his text *Meaning of Relativity* (1922, pp. 55–56)

... from the standpoint of the special theory of relativity we must say, *continuum spatii et temporis est absolutum*. In this latter statement *absolutum* means not only "physically real," but also "independent in its physical properties, having a physical effect, but not itself influenced by physical conditions."

and these absolutes are objectionable since

... it is contrary to the mode of thinking in science to conceive of a thing (the space-time continuum) which acts itself, but which cannot be acted upon.

This theme of a causal defect in earlier theories is a stable feature of Einstein's accounts of general relativity, from his earliest[14] to his latest.[15] The theme was entangled with Einstein's fascination with what became Mach's principles. Yet it survived in his writings after Einstein had abandoned this principle.

Einstein's notion of the causal defect of earlier theories finds clear expression through the work of Anderson. Anderson (1967, pp. 83–87) divides the geometric object fields of a spacetime theory into the absolute $A_1, A_2, \ldots$ and dynamical $D_1, D_2, \ldots$ so that we might write the models as $\langle M, A_1, A_2, \ldots, D_1, D_2, \ldots \rangle$ The absolute objects are, loosely speaking, those that affect other objects but are not in turn affected by other objects. In special relativity, the Minkowski metric would be an absolute object since it determines the inertial trajectories of matter fields, without itself being affected by the matter fields. In general relativity, the space-time metric is dynamical. It fixes the inertial trajectories of matter fields and at the same time its disposition is affected by the mass-energy of the matter field through the gravitational field equations. Because the manifold M together with the absolute objects $A_1, A_2, \ldots$ form an immutable arena, it seems natural to place the cut between spacetime container and matter contained between the absolute objects and the dynamical objects $D_1, D_2, \ldots$

$$\langle M, A_1, A_2, \ldots | D_1, D_2, \ldots \rangle$$

We now have a principled reason for placing the Minkowski metric η_{ab}

of special relativity to the left of the cut and the spacetime metric of general relativity to its right. Anderson's "principle of general invariance" identifies the symmetry group of a spacetime theory with the symmetry group of its absolute structure. This is the group associated with its relativity principle. If the theory has no absolute objects, the symmetry group is the group of symmetries of the manifold itself. Under this principle, the symmetry group of special relativity is the Lorentz group; the symmetry group of general relativity is the general group.

While this analysis offers the most promising explication of Einstein's claims concerning relativity principles, several problems remain. The first is a technical problem. Absolute objects are introduced informally as those objects which act but are not acted upon. Anderson gives a formal definition in which the absolute objects are picked out as those which are the same in all the models of the theory. Friedman (1973) identified the sense of sameness as diffeomorphic equivalence.[16] The definition is too broad for, as pointed out by Geroch (in Friedman 1983, p. 59) all non-vanishing vector fields are diffeomorphically equivalent. Therefore any non-vanishing velocity field in a spacetime theory will be deemed an absolute object. Conversely, as Torretti (1984, p. 285) has pointed out, the definition is too narrow. One can conceive spacetime theories with absolute background structures that fail to be picked out by the criterion of diffeomorphic sameness in all models. For example, one might consider a theory which posits that the background space has some fixed curvature, however the theory does not know what that curvature might be. Its value is located within some range, circumscribed, perhaps, by the reach of observational test. Such a theory would admit models with the relevant curvature drawn anywhere from this range. However the different curvatures of the different models would not be a dynamical response to the amount of matter present in spacetime, as it is in standard relativistic cosmologies. Rather it would merely reflect our ignorance of the correct value of the curvature. The true value of this curvature, whatever it may be, would not vary with a change in the matter content of the spacetime. The background structure would be absolute in the intuitive sense that it acts without being acted upon. However it would not be picked out as absolute by the definition because it fails to be diffeomorphically the same in all the models. Curvature is an invariant property, so structures of different curvature cannot be mapped diffeomorphically onto oneanother. It would seem that the criterion of diffeomorphic sameness in all models falls well short of the intuitive notion of things that act without being acted upon.

A deeper problem is that there remains good reasons for leaving the spacetime metric of general relativity on the left side of the cut. The background spacetime has traditionally been that structure that fixes lengths and times of processes as well as inertial motions. That this structure now responds dynamically to matter does not deprive it of these quintessentially spatio-temporal properties that mark it as belonging to the spacetime container. More seriously, prior to the advent of general relativity and with Einstein's urging, the principle of relativity was understood as expressing an experimental result about the impossibility of distinguishing certain states of motion through space. If we allow that the bare manifold M is the spacetime background, then the principle of relativity ceases to have any direct meaning in terms of experiment. The equivalence of all frames of reference is merely captured in the assertion of the equivalence of suitable congruences of curves with respect to the manifold. There is no direct experimental translation of this equivalence that is akin to special relativity's prediction of failure of all the 19th century aether drift experiments. The principle of relativity of the special theory links the theory directly to its empirical base in experiment. The corresponding principle in general relativity, as it arises in Anderson's analysis, plays no such role. Given all these disanalogies, it is hard to see what in Einstein's general theory ought to be labelled a "generalized principle of relativity", especially if we are interested maintaining some continuity of meaning for the term "principle of relativity".[17]

Finally, Einstein's discussion of the absolutes that act but are not acted upon contains a mysterious element. It was not just that the transition from special to general relativity happened to be accompanied by the elimination of the absolutes. Einstein depicted them as intrinsically defective and demanded their elimination. Anderson (1967, p. 339) expresses this requirement quasi-formally as a "generalized law of action and reaction". The difficulty is to see what compels us to this law. At best one can see loose analogies, perhaps to Newton's third law of motion. However the case of Newton's third law is significantly different. Those who deny it find themselves violating the law of conservation of momentum with all its attendant difficulties. There seems to be no analogous compulsion in the case of Anderson's generalized principle. Newton's mechanics violates the principle without precipitating any obvious problems. Here I agree with Schlick (1920, p. 40) who observed that "Newton's dynamics is quite in order as regards the principle of causality". The problem with Newton's mechanics is that it *happens* to be false. Let us not try to erect a dubious metaphysics merely to convince ourselves that it *has* to be false.[18]

7. CONCLUSION

This modern analysis offers an all too easy diagnosis of Einstein's error concerning relativity and covariance principles. In the Lorentz covariant formulation of special relativity, groups associated with covariance and relativity principles happen to coincide. With the transition to general relativity, the covariance group grew to the general group. What Einstein missed was that the group associated with the principle of relativity did not grow with it. It shrank to the identity group. Had Einstein pursued the geometrical approach of Minkowski rather than his own algebraic approach, he would have been far less likely to confuse covariance and relativity.

This analysis does provide, in my view, a perfectly satisfactory answer to the philosophical question of whether general relativity generalized the principle of relativity of the special theory. As an historical account of Einstein's work, however, it supplies at best a grossly oversimplified caricature. This is already suggested by Anderson's discussion of absolute and dynamical objects. His discussion provides the best modern explication of Einstein's account of the foundations of general relativity and shows how his ideas can be given more precise form. With care, as Stachel (1986, §§5, 6) has shown, Anderson's account can be extended to give precise meaning to Einstein's pronouncement that spacetime cannot exist without the gravitational field.

However another puzzle remains. All this work is focused on taking what Einstein actually said and translating it into a form in which Einstein's original statements are barely discernible. If Einstein's viewpoint was sound, why does it need such dramatic transformation in order for us to see its soundness? I believe there is a better approach that solves this essentially historical puzzle. Einstein's own pronouncements are incoherent to us when read literally only if we fail to take into account the enormous developments in mathematical techniques since the time Einstein wrote. If we account for these changes properly, we find that Einstein can be read literally. His pronouncements on general covariance turn out to be directed at solving a problem peculiar to his simpler mathematical apparatus. This problem remains opaque to us in the modern context since, in part due to Einstein's own efforts, it has been solved automatically and almost completely by the newer mathematical methods. This story is told in Norton (1989, 1992a).[19]

NOTES

[1] See Norton (1993a) for a more detailed survey of the many variant positions that emerged during this debate and how they developed over time as the debate unfolded.

[2] Einstein had decided that no Lorentz covariant gravitation theory could do justice to gravitation. As it turned out, his decision was too hasty. See Norton (1992, 1993b) for Einstein's reasons and the discovery of his error.

[3] Thus Einstein's version of the principle of equivalence is quite distinct in both statement and purpose from the later infinitesimal version of the principle of equivalence which is now universally but incorrectly attributed to Einstein. Einstein's version of the principle did not allow an arbitrary gravitational field to be transformed away infinitesimally. For discussion, see Norton (1985, 1993a, §4.1).

[4] For further discussion, see Norton (1984, §5; 1987) and Howard and Norton (1993).

[5] Kretschmann's footnotes to related literature have been suppressed. The selection of literature cited suggests obliquely that Kretschmann's own earlier work may have been an unacknowledged source for Einstein's point-coincidence argument. See Howard and Norton (1993, §7).

[6] Do we reduce the evaluation to a count of equations and quantities employed? How do we handle alternative formulations of the same theory? How do we count quantities? Is the metric tensor g_{ab} of general relativity one quantity or ten? Is its unique compatible derivative operator ∇_a not to be counted since ∇_a is a derivative quantity fixed once g_{ab} is fixed. Or are we to count g_{ab} and ∇_a as two quantities and add their compatibility relation $\nabla_a g_{bc} = 0$ to the count of equations?

[7] Do Misner, Thorne and Wheeler intend the same when they say "Nature likes . . ."?

[8] Pauli and Weyl's criterion anticipates the later distinction of absolute and dynamical objects. I will remark below on the difficulty of finding a formal characterization of this distinction.

[9] We may even wonder about its success with contrived examples. In the example $A_1 = 0$, the fact of its lack of general covariance means that the equation picks out preferred coordinate systems. Therefore, implicit in the equation is the selection of the relevant coordinate basis vector field u^a. So if $A_1 = 0$ is offered as a physical law, then the basis vector u^a reflects an element of the physical reality depicted by the law and its explicit appearance cannot be ruled out.

[10] I pass over the vagueness of "simplify" in Bergmann's definition. Does it mean "reduce the number of mathematical structures present"?

[11] For this objection, see, for example, Lenard (1921, p. 15) and Einstein (1918a) for the reply.

[12] See for example Earman (1974), Friedman (1973, 1983, Chap. IV), Jones (1981) and Wald (1984, pp. 58, 60, 438).

[13] The correspondence is through the connection between passive coordinate transformations and active boosts. For further discussion, see Norton (1989).

[14] See for example Einstein (1913, pp. 1260–1261). Notice that here as in other places Einstein specifically identifies the inertial system of special relativity as causally objectionable.

[15] See for example Einstein's (1954, pp. 139–140) appendix to Einstein (1922).

[16] Definitions of absolute objects akin to Anderson's but with slight variations are given by Friedman (1973, 1983, pp. 58–60) and Earman (1974, p. 282).

[17] Friedman (1983, Chap. III) has pointed out that we can come close in Newtonian space-time theory in the following sense. Consider those versions of the theory which combine the gravitational field Φ and the flat affine structure ${}^0\nabla_a$ into a single curved affine structure ∇_a. It is possible to decompose this curved ∇_a into many distinct pairs of affine structure ${}^\Phi\nabla_a$ and associated field Φ, all of which are empirically equivalent. Since the various ${}^\Phi\nabla_a$ designate different motions as inertial, one arrives at the kind of extension of the relativity principle that Einstein associated with the principle of equivalence.

[18] I mention briefly the problem of the vagueness of the notion of "acting" in the context of these absolutes. Do universal constants such as Planck's constant h and the gravitational constant G act without being acted upon? One might be tempted to say they are not absolutes in this sense, for there is no physical process connecting h and G with systems in the world. There are no exchanges of energy and momentum, for example. We might wonder, however, whether Einstein would have categorized them as absolutes to be eliminated, for he did hold out the hope for a physics free of arbitrary constants like h and G.

[19] I am grateful to the Research Group: Semantical Aspects of Spacetime Theories (1992/93), Center for Interdisciplinary Research, University of Bielefeld, for helpful discussion.

REFERENCES

Anderson, J. L.: 1967, *Principles of Relativity Physics*, Academic Press, New York.

Bergmann, P. G.: 1942, *Introduction to the Theory of Relativity*, Prentice-Hall, New Jersey; expanded with new appendix, Dover, New York, 1975.

Bondi, H.: 1959, 'Relativity', *Reports on Progress in Physics* **22**, 97–120.

Cartan, E.: 1923, 'Sur les variétés a connexion affine et la théorie de la relativité generalisée',

Annales Scientifique de l'Ecole Normale Supérieure **40**, 325–412, in A. Magnon and A. Ashtekhar (trans.), *On Manifolds with an Affine Connection and the Theory of General Relativity*, Bibliopolis, Naples, 1986.

Cartan, E.: 1924, 'Sur les variétés a connexion affine et la théorie de la relativité generalisée', *Annales Scientifique de l'Ecole Normale Supérieure* **41**, 1–25, in A. Magnon and A. Ashtekhar (trans.), *On Manifolds with an Affine Connection and the Theory of General Relativity*, Bibliopolis, Naples, 1986.

Earman, J.: 1974, 'Covariance, Invariance and the Equivalence of Frames', *Foundations of Physics* **4**, 267–289.

Einstein, A.: 1905, 'Zur Elektrodynamik bewegter Körper', *Annalen der Physik*, **17**, 891–921; translated as 'On the Electrodynamics of Moving Bodies', in H. A. Lorentz et al., *The Principle of Relativity*, Dover, New York, pp. 37–65, 1952.

Einstein, A.: 1907, 'Über das Relativitätsprinzip und die aus demselben gezogenen Folgerungen', *Jahrbuch der Radioaktivität und Elektronik* **4**, 411–462.

Einstein, A.: 1908, 'Berichtigungen zu der Arbeit: Über das Relativitätsprinzip und die aus demselben gezogenen Folgerungen', *Jahrbuch der Radioaktivität und Elektronik* **5**, 98–99.

Einstein, A.: 1911, 'Über den Einfluss der Schwerkraft auf die Ausbreitung des Lichtes', *Annalen der Physik* **35**, 898–908; translated as 'On the Influence of Gravitation on the Propagation of Light, in H. A. Lorentz et al., *The Principle of Relativity*, Dover, New York, pp. 99–108, 1952.

Einstein, A.: 1912a, 'Lichtgeschwindigkeit und Statik des Gravitationsfeldes', *Annalen der Physik* **38**, 355–369.

Einstein, A.: 1912b, 'Zur Theorie des Statischen Gravitationsfeldes', *Annalen der Physik* **38**, 443–458.

Einstein, A.: 1912c, 'Gibt es eine Gravitationswirkung, die der elektrodynamischen Induktionswirkung analog ist?', *Vierteljahrsschrift für gerichtliche Medizin und öffentliches Sanitätswesen* **44**, 37–40.

Einstein, A.: 1913, 'Zum gegenwärtigen Stande des Gravitationsproblems', *Physikalische Zeitschrift* **14**, 1249–1262.

Einstein, A.: 1916, 'Die Grundlage der allgemeinen Relativitätstheorie', *Annalen der Physik* **49**, 769–822, translated without p. 769 as 'The Foundation of the General Theory of Relativity', in H. A. Lorentz et al., *The Principle of Relativity*. Dover, New York, pp. 111–164, 1952.

Einstein, A.: 1918, 'Prinzipielles zur allgemeinen Relativitätstheorie', *Annalen der Physik* **55**, 240–244.

Einstein, A.: 1918a, 'Dialog über Einwande gegen die Relativitätstheorie', *Naturwissenschaften* **6**, 697–702.

Einstein,.A.: 1922, *The Meaning of Relativity*, 5th ed., Princeton University Press, Princeton, New Jersey, 1974.

Einstein, A.: 1933, 'On the Methods of Theoretical Physics', in *Ideas and Opinions*, Bonanza, New York, pp. 270–276.

Einstein, A.: 1940 'The Fundamentals of Theoretical Physics', in *Ideas and Opinions*, Bonanza, New York, pp. 323–335.

Einstein, A.: 1950, 'On the Generalized Theory of Gravitation' in *Ideas and Opinions*, Bonanza, New York, pp. 341–356, 1954.

Einstein, A. and M. Grossmann: 1913, *Entwurf einer verallgemeinerten Relativitätstheorie und einer Theorie der Gravitation*, B. G. Teubner, Leibzig (separatum); with addendum by Einstein in *Zeitschrift für Mathematik und Physik* **63**, 225–261.

Fock, V. (1957) 'Three Lectures on Relativity Theory', *Reviews of Modern Physics* **29**, 325–333.

Fock, V: 1959, *The Theory of Space Time and Gravitation*, in N. Kemmer (trans.), Pergamon Press, New York.

Friedman, M.: 1973, 'Relativity Principles, Absolute Objects and Symmetry Groups', in P. Suppes (ed.), *Space, Time and Geometry*, D. Reidel, Dordrecht, pp. 296–320.

Friedman, M.: 1983, *Foundations of Space-Time Theories*, Princeton University Press, Princeton, New Jersey.

Friedrichs, K: 1927, 'Eine invariante Formulierung des Newtonschen Gravitationsgesetzes und des Grenzüberganges vom Einsteinschen zum Newtonschen Gesetz', *Mathematische Annalen* **98**, 566–575.

Howard D. and J. D. Norton: 1993, 'Out of the Labyrinth? Einstein, Hertz and the Göttingen Answer to the Hole Argument', in J. Earman, M. Janssen and J. D. Norton (eds.), *The Attraction of Gravitation: New Studies in History of General Relativity*, Birkhäuser, Boston, pp. 30–62.

Jones, R.: 1981, 'The Special and General Principles of Relativity', in P. Barker and C. G. Shugart (eds.), *After Einstein*, Memphis State University Press, Tennessee, pp. 159–173.

Kretschmann, E.: 1917, 'Über den physikalischen Sinn der Relativitätspostulat, A. Einsteins neue und seine ursprüngliche Relativitätstheorie', *Annalen der Physik* **53**, 575–614.

Lenard, P.: 1921, 'Über Relativitätsprinzip, Äther. Gravitation 3rd ed., S. Hirzel, Leipzig.

Minkowski, H: 1908, 'Die Grundgleichungen fur die elektromagnetischen Vorgänge in bewegten Körpern', *Königlichen Gesellschaft der Wissenschaften zu Göttingen. Mathematisch-Physikalische Klasse, Nachrichten*, pp. 53–111.

Minkowski, H: 1909, 'Raum und Zeit', *Physikalische Zeitschrift* **10**, 104–111; translated as 'Space and Time', in H. A. Lorentz et al., *Principle of Relativity*, pp. 75–91, 1923; reprinted, Dover, New York, 1952.

Misner, C. W., K. S. Thorne and J. A. Wheeler: 1973, *Gravitation*, Freeman, San Francisco.

Norton, J. D.: 1984, 'How Einstein Found his Field Equations: 1912–1915', *Historical Studies in the Physical Sciences* **14**, 253–316; reprinted in Don Howard and John Stachel (eds.), *Einstein and the History of General Relativity: Einstein Studies*, Vol. 1, Birkhäuser, Boston, pp. 101–159, 1989.

Norton, J. D.: 1985 'What was Einstein's Principle of Equivalence?', *Studies in History and Philosophy of Science* **16**, 203–246; reprinted in Don Howard and John Stachel (eds.), *Einstein and the History of General Relativity: Einstein Studies*, Vol. 1, Birkhäuser, Boston, pp. 3–47, 1989.

Norton, J. D.: 1987, 'Einstein, the Hole Argument and the Reality of Space', in J. Forge (ed.), *Measurement. Realism and Objectivity*, D. Reidel, Dordrecht, pp. 153–188.

Norton, J. D.: 1989, 'Coordinates and Covariance: Einstein's View of Spacetime and the Modern View', *Foundations of Physics* **19**, 1215–1263.

Norton, J. D.: 1992, 'Einstein, Nordström and the Early Demise of Scalar, Lorentz-Covariant Theories of Gravitation', *Archive for the History of Exact Sciences* **45**, 17–94.

Norton, J. D.: 1992a, 'The Physical Content of General Covariance', in J. Eisenstaedt and A. J. Kox (eds.), *Studies in the History of General Relativity, Einstein Studies*, Vol. 3, Birkhäuser, Boston, pp. 281–315.

Norton, J. D.: 1993a, 'General Covariance and the Foundations of General Relativity', *Reports on Progress in Physics* **56**, 791–858.

Norton, J. D.: 1993b, 'Einstein and Nordström: Some Lesser-Known Thought Experiments in Gravitation', in J. Earman, M. Janssen and J. D. Norton (eds.), *The Attraction of Gravitation: New Studies in the History of General Relativity*, Birkhäuser, Boston, pp. 3–29.

Pauli, W.: 1921, 'Relativitätstheorie', in Encyklopädie der mathematischen Wissenschaften, mit Einschluss an ihrer Anwendung, Vol. 5, in Sommerfeld (ed.), *Physik*, Part 2, B. G. Teubner, Leipzig, pp. 539–775, 1904–1922, [Issued 15 November 1921]. English translation in G. Field (trans.), *Theory of Relativity*, with supplementary notes by the author, Pergamon Press, London, 1958; reprinted Dover, New York.

Ricci, G. and T. Levi-Civita: 1901, 'Méthodes de Calcul Différentiel Absolu et leurs Application', *Math. Ann.* **54**, 125–201; reprinted in T. Levi-Civita, 1954, *Opere Matenlatiche*, Vol. 1, Bologna, pp. 479–559.

Schlick, M.: 1920, *Space and Time in Contemporary Physics*, in H. L Brose (trans.), Oxford University Press, New York.

Sesmat, A. 1937, *Systèmes de Référence et Mouvements* (*Physique Relativiste*), Hermann & Cie, Paris.

Sommerfeld, A.: 1910, 'Zur Relativitätstheorie I. Vierdimensionale Vektoralgebra', *Annalen der Physik* **32**, 749–776; 'Zur Relativitätstheorie II. Vierdimensionale Vektoranalysis', *Annalen der Physik* **33**, 649–689.

Stachel, J.: 1980, 'Einstein's Search for General Covariance', paper read at the Ninth International Conference on General Relativity and Gravitation, Jena; reprinted in D. Howard and J. Stachel (eds.), *Einstein and the History of General Relativity: Einstein Studies*, Vol. 1, Birkhäuser, Boston, pp. 63–100, 1989.

Stachel, J.: 1986, 'What can a Physicist Learn from the Discovery of General Relativity?', in R. Ruffini (ed.), *Proceedings of the Fourth Marcel Grossmann Meeting on Recent Developments in General Relativity*, North-Holland, Amsterdam.

Synge, J. L: 1966, 'What is Einstein's Theory of Gravitation', in B. Hoffmann (ed.), *Perspectives in Geometry and Relativity*, Indiana University Press, Bloomington, Indiana, pp. 7–15.

Torretti, R.: 1984, 'Spacetime Physics and the Philosophy of Science', *British Journal for the Philosophy of Science* **35**, 280–292.

Trautman, A.: 1964, *Lectures on General Relativity: Brandeis Summer Institute in Theoretical Physics*, Vol. 1, Prentice Hall, Englewood Cliffs, New Jersey, pp 1–248, 1965.

Wald, R: 1984, *General Relativity*, University of Chicago Press, Chicago.

Weinberg, S.: 1972, *Gravitation and Cosmology: Principles and Applications of the General Theory of Relativity*, John Wiley, New York.

Weyl, H.: 1921, *Space Time Matter*, H. L. Brose (trans.), Dover, New York, 1952.

Department of History and Philosophy of Science
University of Pittsburgh
Pittsburgh, PA 15260
U.S.A.

CARNAP AND WEYL ON THE FOUNDATIONS
OF GEOMETRY AND RELATIVITY THEORY*

At the end of the nineteenth century and, even more, in the early years of the twentieth century the philosophy of geometry experienced unprecedented pressures and tensions. For the revolutionary new developments in the mathematical foundations of geometry and, even more, the application of many of these new mathematical ideas to nature in Einstein's theory of relativity seemed to suggest irresistibly that all earlier attempts to comprehend philosophically the relationship between geometry on the one hand and our experience of nature on the other were radically mistaken. In particular, the Kantian understanding of this relationship – according to which geometry functions as an a priori "transcendental condition" of the possibility of our scientific experience of nature, and space is correspondingly viewed as a "pure form of our sensible intuition" – seemed to be wholly undermined by the new mathematical-physical developments. The question then – for philosophers, mathematicians, and physicists alike – was what new understanding of the relationship between geometry on the one hand and our experience of nature on the other was to be put in its place.

The variety of mutually incompatible answers that were given to this question is remarkable, in that precisely this variety reflects the true complexity – the manifold pressures and tensions – engendered by the radically new philosophical, mathematical, and physical situation. It is especially remarkable, in particular, how seldom a straightforwardly empiricist understanding of the relationship between geometry and experience – according to which geometry is an empirical theory like any other whose validity is straightforwardly verified or falsified by experience – was represented. Pure mathematicians and logicians, for example, who were most concerned to reject the Kantian conception that geometrical reasoning requires "construction in pure intuition" (so that pure geometry is a synthetic science), argued that geometry is first and foremost a purely formal or analytic science having no intrinsic relation to experience whatsoever. Neokantian philosophers, on the other hand, argued that, whereas the new mathematical-physical developments indeed undermined a strictly Kantian conception of the intuitive certainty and experience-constituting character of Euclidean geometry in particular, Kant's most basic and fundamental insight into the "transcendental" function of space and geometry within physics necessarily remained valid. Indeed, even for more physically and empirically oriented thinkers such as Helmholtz it seemed that there must be *some* important respect in which Kant's conception of space and/or geometry as presupposed by empirical physics rather than

127

Erkenntnis **42**: 247–260, 1995.

supported or refuted by empirical physics remains true. It still seemed that there must be something to the idea that we must first ascribe or contribute geometrical structure to nature before we can extract empirical mathematical laws from our experience of nature. Poincaré's conception of geometry as a free stipulation or convention was simply one of the forms in which this fundamental Kantian idea was preserved.

Rudolf Carnap's doctoral dissertation, Carnap (1922), is a particularly interesting attempt to inject order into this rather chaotic situation. The dissertation was written at Jena, where Carnap studied the new mathematical logic with Frege, philosophy with the Neokantian Bruno Bauch, and also experimental and theoretical physics. After first attempting to write a dissertation in the physics department on the axiomatic basis of relativity theory (which the physics department found too philosophical) Carnap ended up receiving his doctorate under Bauch in the philosophy department. In the resulting dissertation Carnap attempts to resolve the contemporary conflicts and tensions in the foundations of geometry – conflicts involving mathematicians, (Neokantian) philosophers, and physicists alike – by carefully distinguishing among three distinct types of space: *formal*, *intuitive*, and *physical* space. Carnap argues that the different parties involved in the various mathematical, philosophical, and physical disputes are in fact referring to different types of space, and, in this way, there is really no contradiction after all: "[all] parties were correct and could have easily been reconciled if clarity had prevailed concerning the three different meanings of space" (1922, p. 64).

Carnap's conclusion is that his third type of space, intuitive space, is synthetic a priori and experience-constituting in precisely Kant's original sense. It is just that we need a more general structure than Kant's original three dimensional Euclidean space:

It has already been explained more than once, from both mathematical and philosophical points of view, that Kant's contention concerning the significance of space for experience is not shaken by the theory of non-Euclidean spaces, but must be transferred from the three dimensional Euclidean structure, which was alone known to him, to a more general structure. However, to the question of what this latter is now to be, the answers are partly indeterminate, in that only isolated characteristics of the three dimensional Euclidean structure are proposed as requiring generalization, and partly contradictory, chiefly because of a failure to distinguish the different meanings of space and insufficient clarity about the conceptual relationship of the space-types themselves – especially the relation of the metrical to the superordinate topological ones. According to the foregoing reflections, the Kantian conception must be accepted. And, indeed, the spatial structure possessing experience-constituting significance (in place of that supposed by Kant) can be precisely specified as topological intuitive space with indefinitely many dimensions. We thereby declare, not only the determinations of this structure, but at the same time those of its form of order in *n*-dimensional topological *formal* space] to be conditions of the possibility of any object of experience whatsoever. (1922, p. 67)

But how are we to understand this remarkable conclusion?

Formal space, for Carnap, is an abstract relational structure whose

initially uninterpreted primitive terms receive a specifically spatial interpretation in intuitive space. And it is here, in fact, that the properly
spatial character of space – that is, its intuitive *spatiality* – is distinguished
from all other relational systems having the same formal structure. This
intuitive interpretation of formal or purely mathematical space is not yet
a physical or empirical interpretation, however. We obtain the latter only
by taking the further step, in Carnap's language, of "subordinating"
[*Unterordnung, Subsumtion*] actually experienced physical phenomena to
the synthetic a priori form of intuitive space. (By contrast, Carnap calls
the more familiar relation between formal and intuitive space "specification" [*Einsetzung, Substitution*].) Carnap sums up the distinction between these two different "application" or "interpretation" relations thus:

> The relation of [formal space] to [intuitive space] is that of the species of structures with
> determinate order-properties but undetermined objects to a structure with these same order-
> properties but determinate objects – viz., intuitively spatial forms. The relation of [intuitive
> space] to [physical space] is that of a form of intuition to a structure with this form made
> up of real objects of experience. (1922, p. 61)

There is no doubt, then, that Carnap intends his notion of intuitive space
to be a generalized interpretation – appropriate to the new mathematical,
philosophical, and physical context – of Kant's conception of space as an
a priori form of intuition.

Carnap explicitly models his notion of spatial intuition on Edmund
Husserl's concept of "essential insight" [*Wesenserschauung*], as developed
especially in Husserl (1913). As explained by Husserl *Wesenserschauung*
functions as follows. Just as in sense perception we are immediately
presented with or immediately given a sensible particular or individual
(a particular color spot, a particular tone, an individual spatial figure),
so in *Wesenserschauung* we can immediately grasp the universal features
that such given sensible particulars exemplify (the general color, the general tone, the general spatial figure). Since *Wesenserschauung* is thus
directed at universal features rather than particular individuals, it is
entirely independent of the particular individuals that actually exist in
the real world. It can, for example, function just as well with an imagined
individual or object of fantasy as with a real or actual individual. Indeed,
for certain purposes, as in pure geometry for example, *Wesenserschauung*
functions even better with purely imagined individuals. And the key
conclusion is now this: since *Wesenserschauung* is independent of the
particular individuals that actually exist in the real empirical world, it is
a source of a priori rather than empirical knowledge. Thus, for example,
through *Wesenserschauung* we obtain a priori knowledge of the structure
of color space, of tone space, or (in pure geometry) of pure geometrical
space. This knowledge is also synthetic a priori rather than analytic a
priori, for it does not follows from the most general truths of pure formal
logic alone. Pure formal logic, for Husserl, expresses the "essence"

[*Wesen, Eidos*] common to all objects of thought whatsoever. The a priori "eidetic sciences" of color, of tone, or of geometrical space, however, hold only for particular subspecies of such objects.

Carnap himself explains the idea as follows:

[H]ere, as Husserl has shown, we are certainly not dealing with facts in the sense of experiential reality, but rather with the essence ("Eidos") of certain data which can already be grasped in its particular nature by being given in a single instance. Thus, just as I can establish in only a single perception – or even mere imagination – of three particular colors, dark green, blue, and red, that the first is by its nature more akin to the second than to the third, so I find by imagining spatial forms that several curves pass through two points, that on each such curve still more points lie, that a simple line-segment, but not a surface-element, is divided in two pieces by any point lying on it, and so on. Because we are not focussing here on the individual fact – shade of color seen here-now – but on its atemporal nature, its "essence," it is important to distinguish this mode of apprehension from intuition in the narrower sense, which is focussed on the fact itself, by calling it "essential insight" [*Wesenserschauung*] (Husserl). In general, however, the term "intuition" may also include essential insight, since it is already used in this wider sense since Kant. (1922, pp. 22–23)

Carnap's debt to Husserl is therefore clear.

But which exactly are the truths of geometry revealed by *Wesenserschauung*? Husserl himself seems to have in mind the totality of truths of Euclidean geometry – as presented in Hilbert's well-known axiomatization, for example. For Carnap, by contrast, the whole problem, as it were, is to adapt Kant's notion of a form of intuition to the general theory of relativity. And here Carnap introduces a most ingenious idea. Euclidean geometry is indeed uniquely presented to us in intuition – more precisely, through Husserlian *Wesenserschauung*. But this very Euclidean intuition is valid only in small or limited spatial regions, so that we are in no way intuitively given, for example, the validity of the parallel postulate in global space. Rather, our intuition tells us only that the Euclidean axioms – which, for Carnap, are definitively give by Hilbert – are satisfied "in the smallest parts" of space in Riemann's sense. Intuition tells us, that is, that space is *infinitesimally* Euclidean; and it is this proposition that expresses the synthetic a priori knowledge characterizing intuitive space.

The point of physical space, for Carnap, is then to order and arrange the objects of our actual experience of nature in the intuitive space we have already constructed completely a priori. This makes sense, for Carnap, because the Husserlian *Wesenserschauung* yielding the basis of our a priori knowledge of intuitive space arises, in the first instance, from our sensible experience of actual spatial natural objects. By considering this experience in its general, essential, or "eidetic" aspects we then arrive at a priori laws governing the structure of intuitive space. Physical space – the space of physical theory – is much more than a mere aggregate of particular intuitively spatial experiences, however; it is rather a precise and consistent ordering of spatial objects in a single mathematical structure. And the point of such a structure is to assign mathematically precise spatial relationships to natural objects – determinate mutual angles and distances, for example – so that precise mathematical laws of nature can then be formulated.

Here, however, a problem arises for Carnap. Intuitive space, as he understands it, has only what he calls topological structure. More precisely, it has only infinitesimally Euclidean structure, so that we are given a priori, as it were, only the entire class of all possible Riemannian manifolds. But the kind of mathematical structure required by physical space is a full (local and global) metrical structure. More precisely, one particular Riemannian manifold must somehow be singled out. How is this to be done? For Carnap, such metrical structure is introduced by a freely stipulated convention. We can, for example, stipulate the Riemannian structure in question directly (as Euclidean, say, in the context of pre-relativistic physics or via the Schwarzschild solution, say, in the context of the general theory of relativity). Alternatively, we can, if we prefer, begin from what Carnap calls a "measure-stipulation" [Maßsetzung] – roughly, the stipulation that a particular physical body is rigid – and then indirectly determine the metrical structure through measurement. The important point, for Carnap, is that, on either alternative, the particular metrical structure one arrives at is in no way inherent in the actual empirical facts of nature; it expresses instead merely the outcome of our own free choice – a choice, to be sure, which is nonetheless absolutely essential, for otherwise we simply could not make precise mathematical determinations at all.

But why exactly is metrical structure thus independent of the actual empirical facts of nature? Carnap explains this idea through the concept of a "matter of fact" [Tatbestand] of experience. We know, from our consideration of the Wesenserschauung underlying intuitive space, that the natural objects actually given to us in experience (or even in mere imagination) have certain necessary or a priori spatial features – just those necessary features expressed in the structure of intuitive space. These – and only these – are the spatial features empirically given natural objects have according to their very nature, as it were. These formal features – and only these – belong to what Carnap calls the "necessary form" of spatial objects of experience. But, as we have seen, the formal features in question are exhausted by what Carnap calls topological structure; no metrical structure (beyond the *infinitesimal* metrical structure) is in fact to be found here. Such further (metrical) spatial structure is thus not inherent in the actual given facts – the "matters of fact" – of experience according to their (spatial) nature, and must instead be conventionally imposed on these facts by us in the form of a freely chosen stipulation:

Now, we have called experience, in so far as it is presented only in the uniquely necessary form that contains no freely chosen stipulations whatsoever, "matter of fact" [Tatbestand]. Therefore, only the spatial determinations contained in matters of fact can be conditions of the possibility of experience. And these, as we have seen, are only the topological, but not the projective and above all not the metrical relations. (1922, p. 65)

Carnap's metrical conventionalism in *Der Raum* is therefore entirely unique and should not be assimilated to any of the other then current forms of conventionalism: that of Dingler, Poincaré, Schlick, or Reichenbach. For Carnap's version rests in the end on his own peculiarly hybrid conception of space as a form of intuition – a form of intuition necessarily lacking a full (local and global) metrical structure.

We can gain a deeper understanding of Carnap's position by juxtaposing it with the in important respects quite similar position developed at roughly the same time by Hermann Weyl. Weyl was of course one of the most penetrating and remarkable contributors to the debate focussed on the new situation in the foundations of geometry and the theory of relativity – especially remarkable because he contributed to the mathematical, to the physical, and to the philosophical aspects of this debate. It appears, moreover, that Weyl's work was centrally important to Carnap's thinking in his dissertation. Weyl (1918) is recommended "in the first place" as a reference for the general theory of relativity, and, for the idea that Euclidean geometry is valid in the small (particularly in the context of general relativity), the reader is referred to Weyl's *Erläuterungen* to Riemann (1919): Carnap (1922, pp. 81, 84). It is also worth noting, in addition, that Carnap (1922) is one of the very few philosophical works in the literature on general relativity cited in Weyl (1922[5]) – along with Schlick (1917) and Cassirer (1921).

The similarities between Carnap's conception and Weyl's are striking indeed. For Weyl also articulates a position according to which space is "essentially" or by its very "nature" *infinitesimally* Euclidean, but not, of course, either *locally* or *globally* Euclidean. Weyl, like Carnap, defends the idea that space is a priori representable by the entire class of Riemannian manifolds. Thus, according to Weyl, what he calls the *nature* of space is given by the circumstance that at each point of the manifold the tangent space bears the same Euclidean metrical structure. What Weyl calls the mutual *orientation* of the metrical structure as we move from point to point, from tangent space to tangent space, is, by contrast, entirely accidental and contingent:

The *nature* of the metric characterizes the a priori essence [*Wesen*] of space in regard to the metrical; it is *one*, and it is therefore also absolutely determined and does not participate in the unavoidable vagueness of that which occupies a variable position in a continuous scale. What is not determined through the essence of space but rather a posteriori – i.e., accidental and capable in itself of free and arbitrary virtual variations – is the mutual *orientation* of the metrics at different points; in reality it stands in causal dependence with matter-and can – participating in the vagueness of continuously variable magnitudes – never be fixed exactly in a rational way, but always only via approximation and also never without the help of immediately intuitive references to reality. One sees that the Riemannian conception does not deny the existence of an a priori element in the structure of space; it is only that the boundary between the a priori and the a posteriori has been shifted. (1918, §13)

And it is clear, moreover, that Weyl, like Carnap, conceives this new

Riemannian conception of the spatial a priori as a generalization of the Kantian conception – a generalization adapted to the new scientific situation created by relativity theory.[1]

The kinship between the two thinkers becomes especially intriguing when we note that Weyl, like Carnap, bears a substantial philosophical debt to Edmund Husserl. Thus, in the Introduction to Weyl (1918), although Weyl explicitly states that he will refrain from entering more deeply into the purely philosophical aspects of the problem in the present book, he nonetheless adds several paragraphs of a self-consciously philosophical nature in which he provides a general outline of the relationship between mathematical-physical theorizing and the realm of immediate subjective experience – the realm of phenomenological "pure consciousness" [reines Bewußtsein]. An endnote to these philosophical paragraphs then explains that "the precise version of these thoughts depends most closely on Husserl's Ideen zu einer reinen Phänomenologie" (1918, note 1 to Introduction). Moreover, at the end of Chapter II of Weyl (1920⁴) – that is, at the conclusion of his own presentation of purely infinitesimal Euclidean geometry as constituting the essence or nature of space – Weyl explicitly connects his analysis with Husserlian phenomenology in an even more striking fashion: "The investigations undertaken in Chapter II concerning space seem to me to be a good example of the essential analysis [Wesensanalyse] striven after by the phenomenological philosophy (Husserl)" (1920⁴, §19). Could it be, then, that Weyl's underlying (if not explicitly expressed) philosophical conception is also close to Carnap's?

That this is not the case emerges when we reflect more closely on how Weyl arrives at the idea that infinitesimally Euclidean geometry expresses the a priori essence or nature space. Does Weyl, like Carnap, arrive at this idea by considering our immediate, quasi-perceptual intuitive consciousness of very small spatial regions? Not at all; he instead presents a subtle and extremely complicated mathematical analysis that is intended to generalize the group-theoretic solution to the "space problem" of Helmholtz and Lie to manifolds of variable curvature, that is, to Riemannian manifolds in general. Just as Helmholtz and Lie derived the Pythagorean (that is, infinitesimally Euclidean) character of the Riemannian metric from a postulate of free mobility – which, however, thereby limited their solution to Riemannian manifolds of constant curvature – Weyl derives the Pythagorean character of the Riemannian metric from a more general group-theoretic argument of a strictly infinitesimal character. Weyl's more general group-theoretic conditions thus yield the Pythagorean (that is, infinitesimally Euclidean) nature of the Riemannian metric without restricting us to homogeneous spaces. The important point for our purposes, however, is that Weyl's deep mathematical analysis of the "space problem" – an analysis he was only able to bring to successful fruition by intensive efforts over several years – has almost nothing in common with Carnap's conception of our a priori knowledge of space.

This stands out particularly clearly, in fact, in the continuation of the passage, cited above, in which Weyl compares his analysis of space to Husserlian *Wesensanalyse*:

The investigations undertaken in Chapter II concerning space seem to me to be a good example of the essential analysis [*Wesensanalyse*] striven after by the phenomenological philosophy (Husserl) – an example that is typical for such cases where it is a matter of nonimmanent essences. We see here in the historical development of the space-problem how difficult it is for us in reality prejudiced humans to discover what is truly decisive. A long mathematical development was required: the great unfolding of geometrical studies from Euclid to Riemann, the physical penetration of nature and its laws since Galileo with all of its ever renewed impetus from the empirical realm, and finally the genius of individual great minds – Newton, Gauss, Riemann, Einstein – all of this was required in order to tear us away from the external, accidental, non-essential characteristics to which we would have otherwise remained attached. Certainly, if the true standpoint is once attained, a light dawns on our reason, and it then knows and recognizes what it can understand out of its own self. Nevertheless, our reason did not have the power (although during the entire development of the problem it of course always "was nearby") to see through the problem in one stroke. This must be held over against the impatience of the philosophers, who believe that they are able adequately to describe the essence on the basis of a single act of exemplary making-present. In principle they are correct, but in human terms incorrect. (1904, §19)

For Weyl, then, the essential structure of space can in no way be discerned "on the basis of a single act of exemplary making-present" [*auf Grund eines einzigen Aktes exemplarischer Vergegenwärtigung*]. On the contrary, only a lengthy historical evolution of mathematical, philosophical and physical ideas can reveal this essential structure to us.

Why then does Weyl, like Carnap, say that his conception of the a priori "essence" of space is based on the Husserlian procedure of *Wesensanalyse*? To understand this we must distinguish two very different ways in which space and geometry may be subject to phenomenological *Wesensanalyse*. One the one hand, geometry – taken merely as a factually given science – may be used to exemplify and motivate the ideas of *Wesenserschauung* and *Wesensanalyse* in the first place. Here we simply take it for granted that a priori geometrical knowledge is based on an intuitive grasp of immediately presented spatial forms (forms that can be intuitively given in fantasy or imagination just as well as in actual spatial perception), and we then use this idea to motivate the possiblity of analogous *Wesenserschauung* in other realms – in the realm of phenomenological "pure consciousness" in particular. This use of geometry is particularly evident in Husserl (1913, §70) where Husserl illustrates the idea that "immediately intuitive grasp of essences . . . can be achieved on the basis of the *mere making-present* of exemplary particulars" [*auf Grund bloßer Vergegenwärtigung von exemplarischen Einzelheiten*] precisely by the example of geometry. It is clearly this way of thinking about geometrical *Wesensanalyse* that inspires Carnap's conception in his dissertation; and it is this way, too, that appears to be the target of Weyl's complaint regarding the "impatience of the philosophers" quoted above.

On the other hand, however, space and geometry may be subject to phenomenological *Wesensanalyse* in a more subtle and, as it were, more constructive fashion. When we make the knowing subject or phenomenological "pure consciousness" itself into an object of *Wesenserschauung* and *Wesensanalyse* we eventually arrive at the point where we consider the phenomenological constitution of the cognitive spatial structure – that is, the phenomenological constitution of *physical* space – in which and through which the knowing subject "looks out" at the physical world. And, as Husserl himself makes clear in (1913, §§136–153), the phenomenological constitution of space in this sense (that is, of physical space) cannot be achieved on the basis of a single intuitive act of *Vergegenwärtigung*. On the contrary, since we are here dealing with the constitution of *reality*, we are necessarily involved with a never-ending approximation to a limiting idea [*Grenzidee*] in the sense of Kant. Essential knowledge of the nature of space in this sense – that is, of the physical space encountered at a definite stage of "transcendental constitution" in the "pure consciousness" of the phenomenological subject – can and indeed must be the outcome of a lengthy constructive process rather than an immediately intuitive act.

This latter kind of approach to a phenomenological *Wesensanalyse* of space – an approach only suggested in Husserl (1913) – was in fact developed in great detail by Husserl's student Oskar Becker in Becker (1923). Becker's idea is to remove the "a priori contingency" that appears to attach to Euclidean geometry considered merely as a factually given "*material* eidetic science" by actually deriving or constructing Euclidean space on the basis of a phenomenological *Wesensanalyse* of the knowing subject. For, if we consider Euclidean geometry merely as a factually given "eidetic science," then, since geometry so understood is not a "*formal* eidetic science" in Husserl's sense (that is, not a branch of pure formal logic), it appears utterly "accidental" and thus "contingent" that Euclidean geometry – as opposed to some other possible geometry – actually describes the essential structure of space. Becker then attempts to remove this "a priori contingency" by constituting specifically Euclidean geometry phenomenologically on the basis of group-theoretic considerations – in an argument that goes roughly as follows: The phenomenological subject is located in a space, in which and through which it perceives the surrounding physical world. The subject must be able to move freely through this space; therefore, by the Helmholtz–Lie theorem, the space must have constant curvature. Moreover, it must be possible to distinguish rotations (by which the subject changes its orientation without changing its position) and translations (by which the subject merely changes its position); therefore, the group of rigid motions must possess a distinguished subgroup of translations, and hence the space must be Euclidean.

Now this phenomenological analysis of Becker's was of course published after Weyl's own group-theoretic work on the "space problem". Nevertheless, it appears likely that Weyl and Becker communicated about these questions much earlier, through their common participation in the pheno-menological circle around Husserl in Gottingen; and, in any case, the two thinkers Becker thanks in his Introduction are precisely Husserl and Weyl. Moreover, it seems clear that Weyl conceives his own analysis of the "space problem" as a generalization and refinement of Becker's – a gener-alization appropriate to the variably curved, non-Euclidean space(-time) of general relativity.[2] For Weyl, too, the phenomenological subject is located in a space, in which and through which it perceives the surrounding physical world. But Weyl does not assume that free motion is possible and thus that this space must have constant curvature. Instead, Weyl begins with the idea of an infinitesimal rotation group at every point – a group that is assumed to be the same (isomorphic) at every point, but is otherwise so far undetermined. We then fix this infinitesimal rotation group as the Euclidean–Pythagorean group by postulating that the thereby induced metric – whatever it is – must determine an associated affine connection *uniquely*. From a phenomenological point of view, then, the subject is located at a "here and now" – about which it must he able to change its orientation. By then postulating that infinitesimal translation (but not necessarily free motion) from this "here and now" is thereby uniquely determined, we guarantee that the space of our subject is *infinitesimally* Euclidean (but not necessarily either locally or globally Euclidean).

In Weyl's later discussion of space and geometry in Weyl (1927) the connection between his own analysis of the "space problem" and the idea of phenomenological constitution is made fully explicit. In a section referring to Husserl (1913) and Becker (1923) – and also, interestingly enough, to Carnap (1922) – Weyl writes:

A way for understanding the Phythagorean nature of the metric expressed in the Euclidean rotation group precisely on the basis of the separation of a priori and a posteriori has been given by the author: Only in the case of this group does the intrinsically accidental quantitat-ive distribution of the metric field uniquely determine in all circumstances (however it may have been formed in the context of its a priori fixed nature) the infinitesimal parallel displacement: the non-rotational progression from a point into the world. This assertion involves a deep mathematical theorem of group theory that I have proved. I believe that this solution of the space-problem plays the same role in the context of the Riemann–Einstein theory that the Helmholtz–Lie solution (Section 14) plays for rigid Euclidean space. Perhaps the postulate of the unique determination of "straight-progression" can be also justified from the requirements of the phenomenological constitution of space; Becker would still like to ground the significance of the Euclidean rotation group for intuitive space on Helmholtz's postulate of free mobility. (1927, §18, pp. 99–100)

This passage strongly confirms the idea that Weyl conceives his group-theoretic analysis of the "space problem" as a generalization and refine-ments of Becker's – an analysis which thereby finds its proper philoso-phical home within the phenomenological theory of the constitution of (physical) space.

Both Carnap and Weyl thus react to the new situation created by the general theory of relativity, not by adopting a straightforwardly empiricist conception of the foundations of geometry, but rather by generalizing the Kantian notion of the synthetic a priori to the *infinitesimally* Euclidean character of space: an intuitive and/or constructive a priori procedure tells us that space is one or another of the infinitely many possible Riemannian manifolds. Moreover, both Carnap and Weyl base their generalization of the synthetic a priori on the Husserlian conception of *Wesensanalyse*. The two thinkers then diverge, however, on how such spatial *Wesensanalyse* is to be understood. Carnap takes (infinitesimally Euclidean) geometry, in Husserlian terms, as simply a factually given "(material) eidetic science" whose essence is revealed to us in immediately intuitive, quasi-perceptual acts of *Vergegenwärtigung* or *Wesenserschauung*. Weyl, on the other hand, explicitly rejects this kind of picture and instead views (infinitesimally Euclidean) geometry as the outcome of a complicated "transcendental constitution" by the phenomenological subject: our synthetic a priori geometrical knowledge is in no way immediately given in mere intuition but rather expresses a phenomenologically based group-theoretic construction requiring all the resources of higher mathematics.[3]

This last idea ultimately leads Weyl to distance himself even further from an immediately intuitive conception of geometry – and thus to distance himself even further from both Carnap (1922) and from Husserl (1913). For Weyl comes more and more to defend the view that, although our mathematical-physical understanding of space must indeed *begin* with immediately intuitive acts (by which the subject locates and orients itself in the "here and now"), the necessary *outcome* of this procedure is a wholly non-intuitive, purely conceptual or "symbolic" construction by which we represent the physical-spatial world by abstract mathematical symbols having no intuitive content. And Weyl comes to emphasize more and more that it is only through such purely symbolic construction, in fact, that we can obtain truly *objective* knowledge of the physical world. This becomes especially clear in a passage from Weyl (1927) where, after mentioning Kant's view of space and time as mere forms of our intuition, Weyl continues:

Intuitive space and intuitive time can therefore not serve as medium in which physics constructs the external world; [we need] rather a four-dimensional continuum in the abstract-arithmetical sense. As colors for Huygens were "in reality" vibrations in the aether, so they now appear only as mathematical functional distributions of a periodic character, whereby four independent variables enter into the functions as representatives of the space-time medium via coordinates. What remains is thus finally a *symbolic construction* in precisely the sense Hilbert carries through in mathematics.

The construction of this objective world, which is only presentable in symbols, from what is immediately given to me in intuition is completed in various *levels*, whereby the progress from level to level is determined by the condition that what is present at one level always reveals itself as an appearance of a higher reality – the reality of the next level. (1927, §17, p. 80)

This conception of a step-wise *objectification* of intuitive experience through ever more abstract purely conceptual construction has much more in common with the Neokantianism of Cassirer – as expressed, for example, in Cassirer (1921) – than with either Carnap (1922) or Husserlian phenomenology.

It is against this backdrop, moreover, that we should understand the striking disagreement between Weyl and Carnap over the conventionality of physical metrical structure.[4] For Weyl himself entirely rejects the idea that physical metrical structure is conventionally stipulated, and instead sees in the general theory of relativity the culmination of Riemann's suggestive remarks according to which the metric of physical space is *empirically* determined by "binding forces" acting on the underlying metrical continuum.[5] This disagreement between Weyl and Carnap should, I think, be understood in the following way. On Weyl's account, space in relativity theory has decisively transcended our spatial intuition. It is rather a purely abstract, purely conceptual structure whose "essence" or "nature" is expressed by the circumstance that the tangent space at each point of the mathematical manifold is Euclidean. It then makes perfectly good sense, for Weyl, to assert that the non-essential, entirely contingent mutual "orientation" of these tangent spaces is empirically determined by Einstein's field equation; for it is Einstein's theory alone that provides us with an adequate *symbolic* construction of objective physical reality. On Carnap's account, by contrast, the a priori infinitesimally Euclidean nature of space is conceived of as a direct reflection of the necessary structure of our spatial intuition. And it follows, for Carnap, that specifically spatial structure (as opposed to purely formal logical-mathematical structure) can only be *intuitively* spatial structure. That this intuitively spatial structure then necessarily lacks what Weyl calls a determinate "orientation" of the purely infinitesimal metrics, can only mean, for Carnap, that no such "orientation" inheres in physically spatial reality at all – in other words, that the full metrical structure of physical space can only be conventional.

Now Carnap, in his later writings, soon leaves the intuitive space of his dissertation completely behind. Indeed, Carnap (1928) adopts a step-wise "constitution" of objectivity via purely conceptual or logical means that is parallel to both the Neokantianism of Cassirer and the symbolic construction of Weyl. Physical space, in particular, becomes a purely abstract mathematical object (the set of quadruples of real numbers $\mathbf{R}^4$) which is distinguished from other isomorphic relational structures only by the circumstance that our previously constituted epistemic subject is perceptually embedded within it – at a definite "point of view," as it were. Indeed, Carnap pushes such logical-conceptual objectification far beyond anything envisioned by Weyl; for Carnap (1928) explicitly asserts that the objectivity of science requires that *all* purely intuitive or ostensive elements must be completely and definitively expunged.[6] Accordingly, Carnap there adopts a strategy of "purely structural definite descriptions" that aims to individuate all objects of science solely on the basis of their formal or structural properties within the logic of *Principia Mathematica*; and it is in this way, in fact, that Carnap finally breaks entirely with both Neokantianism and Husserlian phenomenology. But this is a story for another occasion.

NOTES

* An earlier version of this paper was presented at a workshop on Semantical Aspects of Space-Time Theories at the Zentrum für interdisziplinäre Forschung at the Universität Bielefeld in May 1994. I am indebted to the participants for helpful comments. I am also, and especially, indebted to comments from and discussions with Thomas Ryckman. All translations from the German are my own.

[1] Although Weyl does not explicitly mention Kant in the above-cited passage, closely parallel passages in other works make the Kantian context perfectly clear. See especially Weyl (1922, p. 116) where the infinitesimally Euclidean nature of space is said to be *"characteristic of space as form of appearance"*, and Weyl (1927, §18, p. 97) where Kant is explicitly mentioned.

[2] Becker (1923) also considers the application of non-Euclidean, variably curved spaces in general relativity. According to Becker, however, this application must be conceived instrumentalistically with respect to the uniquely real (Euclidean) space of experience – as a device for simplifying the presentation of physical laws. For Weyl, by contrast, only the variably curved space(-time) of general relativity is the uniquely real or actual space(-time).

[3] Thus, the philosophical passage from the Introduction to Weyl (1918) noted above puts most stress on the idea of physical reality as a limiting idea [*Grenzidee*] – in complete agreement with the conception of physical space sketched in Husserl (1913, §§136–153). These sections from Husserl (1913) – along with Becker (1923) – are in turn approvingly cited in Weyl (1927, §18, p. 92).

[4] This disagreement can be pinpointed very precisely: Carnap (1922, pp. 56–59) uses a particular form of the Schwarzschild metric due to L. Flamm to generate an alternative Euclidean description of the gravitational field involving compensating contractions in our measuring rods – thereby defending conventionalism. Weyl (1918, §33) uses exactly the same form of the Schwarzschild metric to argue that such alternative Euclidean descriptions – although mathematically perfectly possible – are conceptually arbitrary and therefore definitely inferior to the relativistic description.

[5] Compare Weyl's well-known sixth and final *Erläuterung* to the final section of Riemann (1919).

[6] For Weyl, by contrast, the *origin* of our symbolic construction of the objective world in immediately intuitive experience can never be completely overcome, for setting up a coordinate system in the first place necessarily requires an *ostensive* reference to the "here and now". See, for example, the Introduction to Weyl (1918): "But this objectification through the exclusion of the I and its immediate life of intuition does not succeed without remainder; the coordinate system, which can only be indicated by an individual action (and only approximately), remains as the necessary residue of this I-annihilation".

REFERENCES

Becker, Oskar: 1923, 'Beiträge zur phänomenologischen Begründung der Geometrie und ihrer physikalischen Anwendungen', *Jahrbuch für Philosophie und phänomenologische Forschung* **VI**, 385–560.

Carnap, Rudolf: 1922, *Der Raum: Ein Beitrag zur Wissenschaftslehre*, Reuther Richard, Berlin.

Carnap, Rudolf: 1928, *Der logische Aufbau der Welt*, Weltkreis, Berlin.

Cassirer, Ernst: 1921, *Zur Einsteinschen Relativitätstheorie*, Bruno Cassirer, Berlin.

Husserl, Edmund: 1913, *Ideen zu einer reinen Phänomenologie und phänomenologischen Philosophie. Erstes Buch: Allgemeine Einführung in die reine Phänomenologie*, Max Niemeyer, Halle.

Riemann, Bernhard: 1919, *Über die Hypothesen, welche der Geometrie zugrunde liegen*. Neu herausgegeben und erläutert von H. Weyl, Julius Springer, Berlin.

Schlick, Moritz: 1917, *Raum und Zeit in der gegenwärtigen Physik*, Julius Springer, Berlin.

Weyl, Hermann: 1918, 1920[4], 1922[5], *Raum, Zeit, Materie*, Julius Springer, Berlin.

Weyl, Hermann: 1922, 'Die Einzigartigkeit der Pythagoreischen Maßbestimmung', *Mathematische Zeitschrift* **12**, 114–146.

Weyl, Hermann: 1927, *Philosophie der Mathematik und Naturwissenschaft*, R. Oldenbourg, München und Berlin.

Department of History and Philosophy of Science,
Indiana University,
Bloomington, IN 47405,
U.S.A.

ULRICH MAJER

GEOMETRY, INTUITION AND EXPERIENCE:
FROM KANT TO HUSSERL*

In his famous celebratory lecture 'Geometry and Experience' held before the Prussian Academy of Science in Berlin in 1921, Einstein raised the puzzle:

How is it possible that mathematics as a product of human thought, independent of all experience, fits reality so well?[1]

And Einstein immediately offered a short answer to the difficult question he had posed by pointing out firmly that:

Insofar as the sentences of mathematics are related to reality, they are not certain, and insofar as they are certain, they are not related to reality.

This answer is not only typical of Einstein's way of reasoning, it is also in accord with the dogma of Logical Empiricism[2] according to which there are two and only two sources of knowledge, logic and experience. According to this principle, geometrical knowledge – and hence geometry – has to be attributed either to logic or to experience. But this poses a problem, because geometry (as a non-empirical discipline) is usually considered to be a part of mathematics and not a part of science, as Einstein himself (along with many modern physicists) had proposed. Einstein solves the apparently insoluble problem in simple way by splitting geometry into two epistemologically distinct parts: (i) a "pure axiomatic geometry" and (ii) a "practical geometry". The first belongs exclusively to the discipline of *formal logic*, the second is obviously a natural science. "We can regard it as the oldest branch of physics" (Einstein 1921, p. 6).

Einstein, of course, did not rest content with this laconic answer but tried to explain and justify it. In the course of this justification Einstein refers to the new axiomatics, more precisely, to Hilbert's "axiomatic-method". Immediately after the quotation just given, he continues:

Full clarity about the matter, it seems to me, first became widespread through that trend in mathematics, which is known by the name "axiomatics". The progress achieved by axiomatics consists namely in this: through it the *Logical-Formal* became cleanly separated from the material or intuitive content: only the Logical-Formal is, according to axiomatics, the object of mathematics and not the intuitive or any other *content* with which the Logical-Formal is linked.[3]

Although Einstein uses the traditional term "axiomatics", it is obvious that he has Hilbert's *axiomatic method* in mind,[4] because in the succeeding two paragraphs he confronts the *older* with the *newer* interpretation of axioms by giving an example from geometry. This example makes it clear

141

Erkenntnis **42**: 261–285, 1995.
© 1995 *Kluwer Academic Publishers.*

beyond doubt that Einstein is referring to Hilbert's *axiomatic method*, or what is taken to be under this label:

Geometry deals with things, which are denoted by the words "straight line", "point", etc....Knowledge or intuition of these things is not presupposed, but only the validity of th[os]e equally pure formal axioms, ... The axioms first define the things with which the geometry deals. For this reason, Schlick, in his book about epistemology, calls the axioms very strikingly "implicit definitions".[5]

Now, it is not without a certain irony that Einstein refers precisely to Hilbert's axiomatic method when he tries to *justify* his strict separation of geometry into two basically distinct disciplines, because, to my knowledge, Hilbert never made such a distinction between logical-formal and intuitive-contentful geometry. On the contrary, as I will show, Hilbert always adhered to the *unity* of geometry and intuiton or, in other words, to the unity of axiomatic and contentful geometry. To avoid a possible misunderstanding, this does not mean that Hilbert did not make an epistemological distinction between different kinds of geometry, or, more properly speaking between different ways of dealing with geometry.[6] However, Hilbert does not make this distinction in the way Einstein does. Hence something must be wrong with Einstein's appeal to Hilbert, and thus with his understanding of Hilbert's axiomatic method.

In my view, it is Einstein's sympathy with the philosophy of Logical Empiricism which is behind his misinterpretation of Hilbert's axiomatic method. According to the logical empiricists, we have only two sources of knowledge, logic and experience. Unfortunately, geometry does not fit into this clean picture. Hence, what would be more natural than to separate geometry into two aspects (aspects which then do fulfil the requirement of the two sources of knowledge) and to claim that this separation is best captured by Hilbert's axiomatic method. Of course, geometrical expressions like 'point', 'straight-line' etc. must then be taken as *uninterpreted* in order to avoid any question of incompatibility between their mathematical and physical meaning. But this is just the interpretation which logical empiricists had favoured for Hilbert's axiomatic procedure in his 'Festschrift' *Grundlagen der Geometrie* from 1899. And this interpretation is just what Einstein refers to when he tries to justify his distinction between two kinds of geometry. Evidence for this is Einstein's reference to Schlick's "implicit definitions"?[7]

What is so misleading about Einstein's "bifurcated" view of geometry? What bothers me is not the patent circularity in Einstein's justification for splitting geometry into two distinct parts, but rather something quite different:

First and foremost: Einstein overlooks the special character of geometry as a uniform discipline lying precisely between logic and experience. Instead, he deepens the puzzle of the successful application of geometry to reality since the separation of geometry into a pure formal theory and a physical science invokes the question: What has the second to do with the first? The answer, offered by Einstein, that physical geometry is obtained from formal geometry by "coordinating" physical objects to its meaningless terms,[8] is really no answer at all, because the coordination procedure presupposes that we have understood in advance the geometrical meaning of expressions like "rigid body". (Otherwise the question would arise of why, for example, we do not coordinate rubber bands with the formal notion of "rigid body".) In other words, the coordination procedure supposes a solution to precisely the puzzle which has to be solved.

My second objection is more of a fear, namely, that Einstein lends authority to the misinterpretation of Hilbert's axiomatic method by the logical empiricists. This in and of itself is merely an historical accident, perhaps of little interest to the philosopher today. But it becomes *philosophically relevant* if Hilbert's view of geometry itself offers a reasonable way of solving the problem posed by geometry, that is, the problem of explaining its extremely successful application to nature while emphasizing its genuine *mathematical* status. This is precisely, what I want to show in this paper. It is, of course, not an easy task. I will proceed in three stages.

First, I confront the reader with a set of metatheoretical claims made by Hilbert about geometry, claims which look flagrantly incoherent. Nevertheless, I will explain why these claims are not incoherent, contrary to first impressions. This I will do in the second stage by comparing Hilbert's views to those of Kant. Because this is insufficient to resolve the puzzle of applicability, I go one step further and relate Hilbert's views with those of Husserl. This will be done in the third and final section.

HILBERT'S META-GEOMETRICAL CONVICTIONS

When Hilbert took up geometry around 1890 he did this in a very peculiar, unparalleled way, a way which is quite dffferent from the fairytale which logical empiricists later circulated concerning the axiomatic method. Although Hilbert's views about geometry are still in flux in 1891, he has already one fundamental conviction: geometry is, unlike number-theory and algebra, not a branch of logic but rather "the doctrine of the properties of space"[9] and this implied for Hilbert that it rests on intuition and experience:

> The results of these domains (number theory, algebra, function theory) can be achieved by pure thinking.... Geometry, however, is completely different. I can never fathom the properties of space by mere thinking, just as little as I can recognize the basic laws of mechanics, the law of gravitation, or any other physical law in this way. Space is not a product of my thinking, but is rather given to me through my senses. Therefore I require my senses for the establishment of its properties. I require intuition and experiment, just as with the establishment of physical laws.[10]

None the less geometry is a pure mathematical discipline, and in this respect it is distinguished from mechanics and all other natural sciences,

which are, of course, empirical disciplines and for this reason not branches
of pure mathematics. This sounds odd, particularly in the light of the
previous citation. It is, however, an integral part, not only of Hilbert's
early but also his later, more mature convictions. These can be summarized in three statements:

(1) Geometry is the most perfect natural science.[11]
(2) Geometry is (or has become) a pure mathematical science.[12]
(3) The task of geometry is the logical analysis of our spatial intuition.[13]

Taken together, the assertions appear rather incoherent, if not contradictory. But Hilbert maintained consistency among these doctrines by pursu-
ing a philosophy different from Frege's logicism, from Poincaré's geometrical conventionalism, from logical positivism, and last but not least from
neo-Kantianism to which he had some inclinations. But before I outline
what is behind his position, let me mention another triple of distinction
which Hilbert made early on, and which points in the direction, in which
he tried to solve the puzzle of geometry as a science falling between logic
and experience.

In the 19th century, three distinct ways of dealing with geometry co-
existed: (1) *analytical* geometry, (2) *intuitive* geometry (in particular *projective* geometry) and (3) *axiomatic* geometry. This distinction was quite
common, and Hilbert more or less accepted it. This in itself is not very
interesting. What is interesting, however, are the different *roles* or functions which Hilbert assigned to the three types of geometry:

(1)   Analytical geometry is of no significance for the foundations of
        geometry; it presupposes the theory of real numbers as already
        established in advance and reduces then the meaning of geo-
        metrical expressions to analysis through the coordination of
        points with real numbers; its main significance lies in the appli-
        cation of geometry to the (natural) sciences.[14]

(2)   Projective geometry is the most natural and appropriate way
        to represent the facts of our intuition and their complex re-
        lations. But according to Hilbert it lacks the logical clarity
        and conceptual distinctness geometry should entail as the most
        fundamental discipline of all natural sciences.[15]

(3)   The clarity is to be supplied by the third approach, axiomatic
        geometry, as Hilbert calls rather misleadingly the logical analy-
        sis of our spatial intuition and its representation in logical
        deductive form. The significance of the axiomatic approach is
        exclusively of *epistemological* nature: its first goal is to separate
        the *logical* relations from the *descriptive* or intuitional content
        of geometry; its second aim is to arrange the intuitional content
        in such a form that every geometrical fact or phenomena can
        be logically deduced from the axioms.

This is, of course, a very rough and preliminary description of the axiomatic method, but it suffices to indicate the way, in which Hilbert wanted to prove the consistency of his three basic convictions and on this foundation to solve the puzzle of geometry.

The first point to be noted is that, according to Hilbert, the axiomatic method is only *one* approach to geometry among others. This means two things:

(i)   The axiomatic method has no closer relation to geometry than, say, the approach by analytical means. Different approaches to geometry are possible.

(ii)   The axiomatic method has no internal kinship to geometry. In any case it is not restricted to geometry; it can be applied very well to other sciences too.

The last point is particularly useful in grasping why the conflict between the claim that geometry is a natural science and the claim that geometry belongs to pure mathematics is only an apparent one: The axiomatic method is designed particularly to answer one specific epistemological question: Where does the border run between logical and non-logical, contentful thinking? In this respect, however, mathematics (including arithmetic) differs not *in principle* from physics (including geometry): both have a non-logical, intuitive content. (They differ with regard to the question where that content comes from; but this issue cannot be settled by axiomatic means; it must be tackled in a quite different way.) Consequently, neither are pure logical sciences.

But what's with the undeniable difference between mathematics and physics and where does geometry belong, once this difference is clearly established? Here, we must be careful not to make wrong distinctions. Geometry, taken as the science of space in which all external things occur, is doubtless a *natural science*. But it differs from physics by the special circumstance that we have an *implicit* knowledge of all the relevant facts, which belong to geometry. To state the same point more cautiously: In geometry it is extremely improbable (but, of course, not logically impossible) that new facts will come to light, facts which we did not know before. In this and only in this sense geometry is a mathematical discipline in distinction to physics, which depends on new facts brought to light by future experiments.

Given all the facts, it is then the task of the mathematician to analyze and organize the facts in such a way that their logical dependence and independence becomes as clear and distinct as possible. In order to avoid a possible misunderstanding I ought to reiterate the point that the difference between geometry and physics is not a difference in aim. Physics also aims at a clear and distinctive axiomatic presentation. However, in physics we are not as sure as in geometry that we already know all the basic facts: and for this reason the clarity and distinctness to be obtained in physics is not the same as that in geometry. But the difference is only one of degree and not of principle. Let me end this point with the remark that Hilbert's stance was not refuted (as one might suspect) by the rise of special and general relativity. Only the very improbable had happened: completely new and unexpected facts had become known, facts which to some extent also involved geometry.

Until now I have only explained why, according to Hilbert, geometry is at the same time a natural science as well as a mathematical discipline. But I have not explained the *epistemological* reasons for this view. These can't be revealed by the axiomatic method, because they touch the question of the sources of our geometrical knowledge, whereas the axiomatic method takes the facts as given, irrespective of where they come from. Claim (3) expresses Hilbert's basic conviction that our geometrical knowledge is closely connected with our spatial intuition. This claim would be completely harmless under the conditions that Hilbert had meant by "spatial intuition" our empirical impressions of material things in space, and second, had understood by "geometry" any geometry whatsoever. If this were the case, we could choose the 'right' geometry (Euclidean or non-Euclidean), which fits with the facts as we know them from our empirical impressions. In other words, claim (3) would be logically compatible with claim (1) and claim (2); the only difficulty would be to find a reasonable explanation as to why geometry as a mathematical theory fits so extraordinarily well with the empirical data, and in this respect we could fraternize with the logical empiricists. But, of course, we know already that Hilbert does not, for he denies both conditions.[16]

Later writings (to which I will come in the last section) make unmistakeably clear two points: first, that Hilbert means by "intuition" just pure intuition in Kant's *transcendental* sense as a "condition of the possibility of experience", and second that he means exactly Euclidean geometry when he speaks simply of 'geometry' in an epistemological context. This, however, leads inevitably to the question of the *relative consistency* of claim (3) with claims (1) and (2). How can the assertion (3) that it is the task of geometry to analyse our spatial intuition, be reconciled with claims (1) and (2) (that geometry is a physical as well as a mathematical science) if the geometry of our spatial intuition is the Euclidean geometry? This seems to be absurd.

In order to see that Hilbert's position is not absurd but, on the contrary, very reasonable, (indeed much more reasonable than any position I know Weyl's and Carnap's included),[17] we have to go a long way and to compare Hilbert's view on geometry first with the transcendental aesthetic of Kant and second with Husserl's phenomenology. The Kant part will be divided in three steps:

As a kind of prelude I will first compare Hilbert's philosophy of mathematics (Hilbert's label for arithmetic) with that of Kant for the simple reason that, in this terrain, the agreement between both thinkers is maximal. Next, I turn to geometry. Here, the agreement is of a lesser degree, although still remarkable. Yet a certain disagreement occurs, which cannot be removed. In the last section I show how this disagreement forces Hilbert to distance himself in parts from Kant and to move towards a position very similar to Husserl's so called phenomenology, in particular his later conception of "Lebenswelt", regarding the relation of intuition, geometry and experience. This will be investigated more closely in the final section.

HILBERT AND KANT ON ARITITMETIC AND INTUITION

Let me begin with arithmetic. There is not space here to compare Hilbert's position in all its details and ramifications with Kant's, but only in certain points of immediate interest.[18] There are in particular three aspects in respect to which the agreement is almost perfect.

First, both are equally convinced that arithmetic is in need of an intuitive or, as Hilbert also calls it, an extra-logical foundation. For Kant this is a common place; for Hilbert, however, usually dubbed a 'formalist', this has been denied more than once. For this reason I quote a rather long passage from Hilbert's 'New Foundations of Mathematics', which supports my assertion and which re-occurs in almost all publications of Hilbert regarding the foundations of mathematics after 1922:

As we have seen, abstract operations with general extensions of concepts and contents have turned out to be insufficient and unreliable. Instead, as a precondition for the application of logical deductions and the performance of logical operations, something must already be given to us in our faculty of representation: certain extra-logical discrete objects, which are intuitively present as an immediate experience prior to all thinking. To ensure that logical conclusions are reliable, it is necessary that these objects can be surveyed completely in all their parts, and their occurence, their distinction, and their succession is something immediately intuitively present to us in conjunction with the objects as something that cannot be reduced to something else.[19]

Second, it seems that Kant and Hilbert agree that the intuitive foundations of arithmetic are the extra-logical, discrete objects like the 'fingers of my hand' (Kant) or the "strokes on paper" (Hilbert), which can be surveyed completely in all their parts. Kant speaks in this context of 'symbols in concreto',[20] which we use to demonstrate intuitively the validity of an equation like $7 + 5 = 12$ without any appeal to other propositions or concepts. Of course, the similarity of expression does not imply that Hilbert and Kant mean the same when they maintain that arithmetic rests on intuition. Indeed, it is not even quite clear what Kant precisely means by "intuition" in the arithmetical context. But one negative demarcation can be stated, which shows how closely related Kant's and Hilbert's views indeed are.

Kant does not mean, as is usually stated, that time is *the* characteristic form of intuition in arithmetic. This opinion, which has been propagated

mainly by Brouwer and his intuitonistic school, is to my mind not correct
and has been refuted by Kant himself. Time is, according to Kant, only
typical for the phenomena of motion (or change in general) but not for
arithmetic. "Pure mathematics", says Kant in his dissertation, "considers
space in geometry, time in pure mechanics".[21] Although, counting takes
place *in* time, like any other human activity, this fact in itself is no
sufficient reason to turn arithmetic into a temporal science. In geometry,
too, the drawing of a line takes time, and yet the characteristic form of
intuition in geometry is space and not time. Essential for arithmetic as
the doctrine of magnitudes is exclusively the reference to discrete singular
objects like the fingers of my hand or the beats of a clock; whether they
are space-like or time-like is irrelevant. Precisely in this point, Hilbert
and Kant agree: the epistemological foundation for arithmetic is not time
as such but its reference to extralogical objects. Of course, these objects
only occur either in space, or in time, or in the union of both. (This, by
the way, is the reason why they are extralogical objects.) Hence, arithmetic
is not possible without at least one of these forms of intuition, although
which one is a contingent matter.

Third, Hilbert and Kant agree that arithmetic begins with (or perhaps
better, has its origin in) absolutely concrete, singular propositions like
"$2 + 3 = 5$" or "$3 > 2$" and not with general propositions or laws, as
indeed most of the other sciences do. This agreement is particularly remarkable, because it shows that Kant and Hilbert share a certain peculiar
view of arithmetic, which as far as I know nobody else affirms. What
follows from this idiosyncratic view?

According to Hilbert, the propositions involved express "matters of
fact" and not e.g., logical truths or conceptual relations. Viewed in this
way, arithmetic originates from concrete facts and not from universal laws.
This is extremely important for the foundations of arithmetic, because it
permits us to distinguish two different domains of arithmetical propositions: (i) the domain of singular propositions expressing the simple facts,
and (ii) the domain of universal and existential propositions about numbers and their properties and relations. The first domain is unproblematic
in the twofold sense that its propositions obey the principle of excluded
middle and are, taken together, obviously consistent. The second domain
is, in contrast to the first, more problematic, because its propositions
express no facts, are neither true nor false, but *ideal* assumptions, whose
sole purpose is to explain, via deduction, and to unify and simplify the
propositions of the first domain.[22] For this reason, they cannot be asserted
as true, but only assumed hypothetically as *axioms*, and their internal
consistency has to be proved separately. In other terms, the epistemological status and function of both types of propositions, of singular propositions and axioms, is basically different. It is precisely this difference, which
Hilbert has in mind, when he says:

When number theory is carried out in this [intuitive] way, there are no axioms, and so no contradictions of any kind are possible. Indeed, we have concrete signs as objects, we operate with them and make content-ful statements about them.... But certain, the whole of mathematics can-not be presented in this way. Already by the transition to higher arithmetic and algebra, for example, when we want to arrive at assertions about infinitely many numbers or functions, this contentual way of proceeding fails. Since, for infinitely many numbers we cannot write down number signs or introduce [appropriate] abbreviations.[23]

This means that, first in the second infinite domain of arithmetical propositions, axioms become *essential*, because we cannot talk about infinite totalities of numbers or functions without introducing some ideal assumptions like the validity of the "tertium non datur" or general and existential claims.[24] In the restricted domain, however, in which we deal only with intuitively given, finite sets of concrete numbers, the axioms are only *means* of simplifying and unifying the set of singular propositions and can in principle be dismissed. In his later work, *The Logical Foundations of Mathematics*, Hilbert explains more closely his view regarding the relation between the first and second domain, in particular, why the axioms with the ideal assumptions can be dismissed, at least in principle, in favour of the intuitively given finite domain of numbers:

By means of the axioms 1. to 10. we obtain easily all the positive integers and the numerical equations, which hold for these. Also, through this beginning, we can achieve elementary number theory by means of 'finite' logic through purely intuitive considerations to which belong recursion and intuitive induction for the finite totalities with which we are presented: without utilizing any objectionable or problematic inferential procedure. The provable formulas, which can be achieved by this standpoint, all have the character of the finite, i.e. the thoughts, whose representations they are, can also be obtained by contentually and immediate consideration of finite totalities *without any axioms*.[25]

In the intuitively given finite domain of singular propositions about numbers the role of the axioms is obviously only that of a *tool*, but, of course, an extremely useful tool, which it is worthwhile to study for its own sake. That Kant with respect to arithmetic advocates essentially the same view, I cannot develop here in all its necessary details. I can, however, quote two passages, which confirm my claim. The first passage is from the *Critique*. In Chapter II about the 'System of all Principles of Pure Understanding', Kant argues (in the context of the Axioms of Intuiton) with respect to propositions about numbers (*quantitas*) and their relation to axioms as follows:

As regards magnitude (*quantitas*), that is, as regards the answer to be given to the question, "What is the magnitude of a thing?" there are no axioms in the strict meaning of the term: although there are a number of propositions which are synthetic and immediately certain (*indemonstrabilia*). The propositions, that if equals be added to equals the wholes are equal, and if equals be taken from equals the remainders are equal, are analytic propositions. [Consequently, they are not] axioms, [for these] have to be *a priori* synthetic propositions. On the other hand, the evident propositions of numerical relation are indeed synthetic, but are not general like those of geometry, and cannot, therefore, be called axioms but only numerical formulas. The assertion that 7 + 5 is equal to 12 is not an analytic proposition. But although the proposition is synthetic, it is also only singular.[26]

There is a second, quite similar remark in a letter to the mathematician Johann Schulz of 25 November 1788, which shows that Kant was not muddled, but on the contrary quite clear about the essential point that finite arithmetic needs no axioms:

Arithmetic certainly has no *axioms*, because, properly speaking, it has no *quantum* – i.e., no object of intuition as magnitude – for its object, but merely quantity, i.e. a concept of a thing in general through the determination of magnitude. It has, however, postulates, i.e. immediately certain practical judgements . . . . The judgement $3 + 4 = 7$ seems to be merely a theoretical judgement . . . the $+$[denotes] a kind of synthesis through which a third number is to be found out of two given ones, a task which is neither in need of a solution procedure nor of a proof. Hence the judgement is a postulate.[27]

Although the difference between quantum and quantitas is difficult to grasp and requires detailed commentary, the main point, the non-axiomatic character of elementary number theory, is clear, I think. Therefore let us turn immediately to geometry.

INTUITION AND GEOMETRY

Regarding geometry, one should expect a basically different relation between Hilbert and Kant than in the case of arithmetic. Whereas in the latter case, the agreement can be, and indeed is, almost perfect because in arithmetic no dramatic revolutions had intervened since Kant's time, the situation in geometry is completely different. Kant's view that geometry is based on pure intuition seems to be refuted once and for all by the emergence of non-Euclidean geometries during 19th century and by their successful application to nature in the theories of special and general relativity. In any case thus or very similar argues M. Friedman in his recent book *Kant and the Exact Sciences*.[28] It is surprising, therefore, to hear that Hilbert himself did not share this opinion, that, on the contrary, he conceded a certain legitimacy to Kant's transcendental philosophy not only in regard to arithmetic but also with respect to geometry, and that precisely in spite of the scientific developments just referred to. This does not mean, of course, that Hilbert agreed with Kant in every respect: on the contrary, in the case of geometry the differences are greater and more fundamental than in the case of arithmetic. Let me first explain the two most fundamental agreements and then stress the differences.

(1) The most important agreement is, by far, the common conviction that geometry is based on intuition. This is not only explicitly stated by Hilbert at the beginning of his 'Festschrift' (and all the other early lectures) about the foundations of geometry,[29] but also stressed by Hilbert in later years as well, after relativity theory had been established. He opens the 'Festschrift' with a quotation from Kant's *Critique*, which, to my mind, has to be taken very seriously, indeed literally, because it expresses, as I will show, Hilbert's epistemological point of view very accurately, and this not only with respect to geometry but to all theoretical sciences:

So fängt denn alle menschliche Erkenntnis mit Anschauung an, geht von da zu Begriffen und endigt mit Ideen. [A702/B730][30]

And immediately in the very first paragraph of the Introduction Hilbert says: "The designated task amounts to the logical analysis of our spatial intuition". But as late as 1930 Hilbert in his essay 'Naturerkennen und Logik' emphasized his basic agreement with Kant:

Thus the most general and fundamental idea of Kantian epistemology retains its significance: namely, the philosophical problem of determining that intuitive attitude a priori and thereby of investigating the conditions of the possibility of all conceptual knowledge and of all experience.[31]

In view of this and numerous similar passages, I think it is impossible to dismiss Hilbert's reference to Kant simply as a *façon de parler* or to pass over it in silence as most philosophers and mathematicians now do. On the contrary, it would be important to know how Hilbert arrived at this opinion and even more important to understand how it was possible for him to hold this view without being in contradiction with the development of modern geometry and its successful application in physics. But before I take up this challenge, I would like to discuss a further common ground between Hilbert and Kant.

(2) In his (1918) lecture on *Space and Time*, Hilbert, referring to the relationship between geometry and space, states clearly and unequivocally:

The exact systematic collection of the properties of space, the investigation of the logical relations between them and the development of the consequences which result from them form the [proper] content of Euclidean geometry.[32]

At first glance, this in fact is a rather surprising statement from Hilbert, since Euclidean geometry seems to be granted a *peculiar* status. However, for the moment, I would like to leave it at this and reserve the solution of this riddle for the conclusion. This statement interests me here for a different reason.

If you take this statement together with the statement about spatial intuition in the 'Festschrift', then this means: Euclidean geometry is *the* theory of space, its properties and logical relations, as we "intuitively" grasp it respectively as we represent it in our sensible imagination, that is, as a system of possible relations between things, in particular their shape and relative positions.[33] But Kant determines the relation between geometry, space and external intuition in nearly the same way. In *The Transcendental Exposition of the Concept of Space* he says: "Geometry is a science which determines the properties of space synthetically, and yet a priori. What, then, must be our representation of space in order that such knowledge of it may be possible?" And the answer is: "It must in its origin be intuition; for from a mere concept no propositions can be obtained which go beyond the concept – as happens in geometry" [*Critique*

B40/41]. This means that, according to Kant, Euclidean geometry is a *synthetic* a priori science of space precisely because it is the theory of our external, object-related intuition. This in turn is nothing other than our representation of space as a system of things qua *external appearance*. If one ignores for the moment the possible difference between space "as a system of things qua external appearances" and space as a "system of the possible relative positions of things", then the epistemological relation between geometry, intuition and space for both Hilbert and Kant is exactly the same: Euclidean geometry is the theory of our external intuition and this in turn is simply our representation of space as a system X, be it "the things qua external appearances", be it "the possible relative positions of things".

Thus it seems, as if I would like to suggest that Hilbert intended a "return to Kant". But that is not the case. In order to convince the reader, I simply have to explain the remaining fundamental difference between Hilberts and Kant's interpretation of geometry as the science of space respectively of our spacial intuition. I will do this in two steps which appear to be rather different but which in fact are closely related.

(3a) For Kant, geometry – quite unlike arithmetic – is in essence an *axiomatic* science, which means that it proceeds from *general statements* or principles instead of concrete immediately certain judgements or postulates as Kant names them. From these general principles we derive singular propositions by way of 'constructing' the geometrical concepts in (pure) intuition. For Hilbert, geometry is different. Hilbert views geometry epistemologically as closely related to arithmetic: both are based on *facts*, indeed on facts of our finite intuition, i.e., singular judgements about "intuitively given, extra-logical objects", like "straight line *a* is twice as long as straight line *b*" and similar statements. In this sense Hilbert remarks in his 'Festschrift':

We can divide the axioms of geometry into five groups; each one of these groups expresses certain interrelated fundamental facts of our intuition.[34]

At first glance, this quotation seems to contradict my interpretation that, according to Hilbert, geometry is founded on facts of our intuition, because Hilbert obviously starts from axioms and not from singular propositions as in arithmetic. The apparent contradiction is, however, easily resolved: In arithmetic, too, we have axioms, but only as an "instrument" while we remain in the finite. Hence, if we transfer the relationship between facts and axioms in arithmetic to geometry, then we can, at least in principle, do without axioms in geometry, as long as we remain in an intuitive-finite domain. This in turn means that the axiomatic representation first becomes essential in geometry when one transcends the finite and steps into infinity. This, unquestionably, does take place in Euclidean geometry. The infinity is, however, in both cases, arithmetic as well as Euclidean geometry, the result of an "idealisation" or, as Hilbert puts it:

merely [an] *idea* – if one means by an idea, in Kant's terminology, a concept of reason, which transcends all experience and by which the concrete becomes completed in the sense of a totality".[35]

This means that in geometry, as in arithmetic, the axiomatic representation only serves as a means to simplify and unify the relations in the finite by including the regulative idea of infinity. If this interpretation is correct, then it implies a number of additional, important differences between Hilberts and Kant's views on geometry. For now, I will deal only with one difference, that which is most fundamental for the further course of the argumentation.

(3b) If it is correct that both Hilbert and Kant share the view that Euclidean geometry is an immediate expression of our spatial intuition, then this in turn has the unexpected consequence, based on Hilbert's argument for the ideality of Euclidean geometry, that our outer intuition in itself already involves ideal elements. And this is exactly what Hilbert maintains, because although space is *unlimited*, it is not *infinite* (at least not infinite because it is unlimited) for the simple reason that the two concepts are *logically independent*, as Hilbert points out: "Unlimitedness and finiteness do not exclude each other, as can be seen from the example of a two-dimensional spherical surface".[36] On the contrary, the assertion of the infinity of space is a genuine conceptual *extension* of the assertion of its unlimitedness. This view brings Hilbert for the first time into genuine conflict with Kant. More precisely, Hilbert thinks that Kants conclusion from the unlimitedness of space to its infinity is a logical mistake:

Up to the time of Kant and beyond, there was no doubt about the infinity of space. But this opinion rests on an error in thinking. From the fact that beyond a segment of space, always more space is present only the unlimitedness of space can be [logically] deduced but definitely not its infinity.[37]

This confronts us with the difficult question: Why does Hilbert retain Kant's view that Euclidean geometry is the only correct or appropriate theory of our intuition of space when, at the same time, he criticizes Kant's view of space to rest on a serious error of logical reasoning. This seems to be rather inconsequential! For a long time, I had no answer to this question. But now I believe that I have found the correct answer, at least in principle. In the final chapter I will explain the outlines of the answer.

GEOMETRY AND THE THEORY OF RELATIVITY:
HILBERT'S MOVE TO HUSSERL

The core of the answer can be found in two lecture series on the genesis and development of the *special* and *general theory of relativity*, which Hilbert held in the winter term 1918/19 and the summer term 1921. The first is his lecture on 'Space and Time' [S&T], already mentioned, and the second his lecture about 'Grundgedanken der Relativitätstheorie' [GGRT], which, although two years apart, are very closely related in content. The first develops the objective, experimental reasons for the transition from the classical conceptions of space and time first to the special and then to the general relativistic conception of spacetime. The second reflects the *necessity* of this development in a more philosophical manner, a manner which is of particular significance here. Hilbert's view, in brief, is this:

There are *two* conceptions of space-time, a traditional one, rooted in the "Lebenswelt", and a scientific view which is coupled to the development of physics. I chose the term "Lebenswelt" quite deliberately because Hilbert means something similar to what Husserl describes with this expression in his book *Die Krisis der Europäischen Wissenschaften und die Transzendentale Phänomenologie*, when he distinguishes between the common conception of space and time of the *everyday life* and the *scientific* conception of space and time, in particular the four-dimensional spacetime of the theory of relativity.

In his lectures Hilbert again and again emphasizes that our classical conception of space and time as characterized by Euclidean geometry and the Galileo-group of transformations is, in the practical sense, fully correct, and not only that, but even *after* the development of the theory of relativity, the classical conception is the accepted basis of all our *practical* actions. Morever this conception was in a certain way (which will be characterized more closely in a moment) the epistemic *presupposition* for the development of the theory of relativity, without which the detection of facts which finally forced scientists to create and adopt the theory of relativity would have been quite impossible. To support my claim let me quote only two of many sections from the GGRT:

Thus we have listed all the essential features of the old conception of space, time and motion. But before we turn to our second question, it is still absolutely necessary to bring to mind how excellent this conception of space-time has proved to be. As far as natural sciences and their applications are concerned, we find that everything is based on this conception. And in this construction everything fits together perfectly. Even the boldest speculations of physicists and astronomers are brilliantly confirmed in the minutest detail so that one can say that the experiences of everyday life, our practice and custom, the traditional intuition and the most demanding sciences were in complete agreement and most pleasant harmony with each other. (p. 20)

This prevailing conception of space-time was the framework within which our entire knowledge of nature, in particular of mechanics, physics and astronomy, this rich, expansive, manifold and diversely specified material fits perfectly. And what is more, this conception of space-time did not develop from science and not at all from reflection. On the contrary we use the concepts space and time in *daily life* in this way and experience them continually anew so that they are familiar to us and we feel comfortable with them.

So it happened that the laws of geometry and the geometrical theory of motion in which this conception of space-time finds its precise expression and which represents the permanent tools, the ABC of the physicist so to speak, were simultaneously viewed as something preceding all physics which can be recognized without experiment by intuitive reasoning.

Thus the theory of the sum of the angles in a triangle or the theory of the addition of velocities are just as valid and with the same degree of certainty as the laws of arithmetic. (GGRT pp. 30–31)[38]

I think, these quotations testify that Hilbert's view regarding classical space-time theory indeed is closely related to Husserl's concept of "Lebenswelt", especially since Husserl too conceives of space-time, according to which we act in daily life, as being for the most part independent of the historical development of the sciences. Since it existed long *before* the development of modern science, and it remained quite exactly the same, even *after* the development of the theory of relativity. So far, so good.

Yet, a trivial objection could be raised to this interpretation: namely, Husserl wrote *The Crisis of the European Sciences* in the mid-thirties whereas Hilbert gave his lectures in 1918 and 1921 respectively. Hence, Hilbert could not have known Husserl's conception of "Lebenswelt". But the objection is no objection at all, because in the first place Hilbert *could* have developed his particular view quite independently of Husserl. In this case we would have a mutually independent creation of related epistemological views, because it is almost certain that Husserl did not know Hilbert's lectures on relativity.[39] However, the situation is actually more interesting. Not only had Husserl already developed the nucleus of his later distinction between science and 'Lebenswelt' in 1905 in his 'Five Lectures about The Idea of Phenomenology', but he had developed it probably in full awarness of most of Hilbert's work on geometry.[40] And this last point is the crucial one, because (as we now know) Hilbert made the distinction in question already in his early work on geometry. In his first lecture series on geometry in 1891, as a 'Privatdozent' in Königsberg, Hilbert already remarked: "Indeed, the oldest geometry[41] also arises from the intuition of things in space, as it is offered up by daily life, and, like every science in its beginnings, had posed problems of a practical nature".[42] Of course, this might be seen as a sloppy remark, but it is not, and on the contrary it is intended as a serious claim that the epistemic origin of Euclidean geometry becomes evident from the fact that Hilbert repeats it at the beginning of his lectures on 'The Foundations of Euclidean Geometry' in 1898. Here he says: "This [Euclidean] geometry is in a way the geometry of daily life. It is the basis of all consideration of nature and natural science".[43]

It is not my intention to maintain that Husserl was indeed influenced by Hilbert's view in this respect. Although we know that Husserl had close contact with Hilbert in the years shortly before and after the turn of the century (it was Hilbert who had initiated Husserl's call to Göttingen) I will leave aside the thorny question of who influenced whom. Instead I want to point out a striking similarity between Husserl's and Hilbert's epistemological point of view with respect to mathematics as whole, not only of geometry but also of arithmetic. Both were *finitists* with respect to the question of where the certainty of our mathematical knowledge comes from. And yet, at the same time, both believed in *Cantor's para-dise*, that is, in the legitimacy of the (actual) *infinite*. This attitude posed for both thinkers a serious problem, namely: how can the *transition* from the finite to the infinite be justified? In Hilbert's terms: how can the infinite be justified on the basis of the finite? In spite of this common problem it is not unlikely that both thinkers finally came up with a similar solution. In fact, in Husserl's case we have ample evidence that his late distinction between the sciences and the *Lebenswelt* was intended to solve precisely this problem. But what does the solution look like?

To answer this question is in Husserl's case not an easy task, particularly not with respect to geometry, because Husserl himself never published a detailed proposal how Euclidean geometry, which obviously takes space as infinite, is constituted in a phenomenological manner on the basis of our *finite* intuition rooted in *Lebenswelt*. Aside from some suggestions in *Ding und Raum* (1907) and *Ideen* (1913), he seems to have delegated the tricky task to his students E. Stein and O. Becker. The best I can do in this respect, is to refer the reader to Becker's excellent work 'Beiträge zur phänemenologischen Begründung der Geometrie und ihrer physikalischen Anwendungen' (1922). Yet one qualifying remark is required. However sceptical the reader may be about Becker's phenomenological justification of Euclidean geometry (a scepticism, which I share because the so-called *Wesensanalyse* of the phenomenal space relies too much on psycho-physical considerations about the subject apprehending space) the reader should not underestimate Becker's phenomenological approach as epistemologically naive or inferior to Weyl's *infinitesimal* analysis of space, as M. Friedman suggests.[44] Two simple reasons can be raised against this estimation. In the first place, Weyl's "infinitesimal" analysis of space is itself problematic, both from a physical perspective and from Hilbert's axiomatic point of view, because it presupposes the validity of Euclidean geometry in the infinitesimal small and, hence, takes as granted what is, properly speaking, the task of the axiomatic approach, namely an investigation of the dependence and independence of the different axioms which together amount to Euclidean geometry. Second, and much more important, there is a different reading of the phenomenological constitution of Euclidean space, which makes perfect good sense from a physical as well as methodological point of view, and this is Hilbert's "transcendental" approach to Euclidean geometry, to which I now turn.

To begin with, I have to answer the question: in what sense is the conception of space-time as rooted in the *Lebenswelt* a presupposition for the development of the special theory of relativity?[46] For the sake of brevity, I can only indicate the direction, in which the correct answer lies.[47] Only on the basis of the traditional conception of space-time was it possible to increase the precision of experimental arrangements and measuring methods to such an extent that one could prove by means of very sophisticated optical experiments that Galileo's theorem for the addition of velocities is *false*, strictly speaking, or at least only approximatively valid. Yet, because the addition theorem is an immediate logical consequence of the classical conception of space-time, hence, this too had to be rejected and a far more general theory put in its place: the special theory of relativity. (The adjective 'special' is very misleading because the *principle of relativity* as formulated in the special theory of relativity covers *many more phenomena* than Galileo's corresponding principle in classical mechanics. The latter was defined only for macroscopic bodies like ships and planets but not for light-rays, whereas the principle of relativity in the "special" theory covers also the propagation of light (as well as other electromagnetic waves).)

Two opposite objections can be raised against this "dialectical" procedure: The first says that one cannot presuppose *false* assumptions like the traditional conception of space-time, for this is to presuppose nothing at all. The second maintains that the procedure as characterized by Hilbert is illogical, or even worse, that it entails a *vicious circle* because it destroys the assumptions (of classical space and time), on which basis the experiments and measurements were performed. Consequently, classical space-time cannot simply be suspended and another theory, special relativity, put into its place. Both objections are unconvincing. First, physicists have certainly used classical space-time conceptions; indeed, all experiments until recently were designed and performed on this basis. Second, there is just no serious circularity or illogicality in the procedure indicated. On the contrary it is a paradigm of logical reasoning, and therefore I suspect that the error is on the side of the opponents; probably they were mislead by the *several* meanings of the terms 'assumption' and 'presuppose'. But, whatever the error is, I will concede this much: the logical meaning of the term 'presupposition' has not been sufficiently analysed and logically explained: In what sense does the special theory of relativity presuppose the classical conception of space-time and, hence, of Euclidean geometry?

In order to answer this question, I will compare Hilbert's conception of "presupposition" with Kant's transcendental conception of Euclidean space as one of the two aesthetic conditions of the possibility of experience. In particular I will ask: how compatible is the distinction between *two* conceptions of space (one anchored in the *Lebenswelt*, the other the result of 'scientific' investigations) with Kants *undivided* conception of space as one of the forms of pure intuition which is at the same time, and this is quite essential for Kant's transcendental consideration, one of the "conditions for the possibility of experience"? The answer bears certain "dialectical" features in that it confirms Kant's conception in one respect, yet contradicts it in another, and both together are combined in a third phenomenologically modified position.

First, the positive aspect of distinction which confirms Kant: For Kant, Euclidean geometry was a synthetic a priori science of space exactly because space is only the *subjective* form of external intuition. The latter

for Kant, however, means nothing less than that space is a formal character
of the external senses and determines how the subject is "to be affected
by the objects and thus obtaining an immediate representation, that is,
intuition of them" [B 41]. Thus it is completely consequential when Kant
says: "The constant form of this receptivity: which we term sensibility, is
a necessray condition of all the relations in which objects can be intuited
as outside us; and if we abstract from these objects, it is a pure intuition,
and bears the name of space" [B43/A27].

This subjective interpretation of space as a form of representation or
intuition of external objects is taken up by Hilbert in the distinction
between "*lebensweltlicher*" and scientific conceptions of space (viz., of
space-time) when he, like Kant, supposes that the conception of spacetime, which is rooted in the *Lebenswelt* is a kind of *anthropological constant* which is, for the most part, independent of respective developments
in science. Just as today we still distinguish between "up" and "down",
in spite of the Copernican revolution, we also retain the classical conception of space-time in everyday life despite the acceptance of the theory of
relativity. The anthropological constraint on the everyday conception of
space-time is, however, for Hilbert precisely the prime reason why he
distances himself from Kant in one very specific aspect.

Whereas for Kant the *subjective* forms of sensibility are simultaneously
the *objective* conditions of the perception of objects and thus the central
conditions for the possibility of experience, for Hilbert this is definitely
not the case. On the contrary, the goal of science is to free our cognition
from the mere subjective conditions of sensibility and to try to attain
an *objective*, observer independent cognition. Hilbert includes, however,
among the subjective conditions of sensibility not only the secondary
qualities such as colours, sounds and smells, but also space and time as
the two forms of human intuition rooted in the *Lebenswelt*. This is the
deeper reason why Hilbert differentiates between *science* and *Lebenswelt*,
between a *everyday* conception of space and time and only views the
former, not the latter, as objective:

Hitherto, the objectification of our view of the processes in nature took place by emancipation
from the subjectivity of human sensations. But a more far reaching objectification is necessary, to be obtained by emancipating ourselves from the *subjective* moments of human
intuition with respect to space and time. This emancipation, which is at the same time the
high-point of scientific objectification, is achieved in Einstein's theory, it means a radical
elimination of *anthropomorphic* slag, and leads us to that kind of description of nature which
is *independent* of our senses and intuition and is directed purely to the goals of objectivity
and systematic unity.[48]

Although this statement is clear and quite unmistakable in its insistence
on separating the *subjective* intuiton of space and time from their *objective*
structure (as represented in our best physical theories), two questions
must to be raised which are not answered by Hilbert's statement itself.
First, what does Hilbert understand by space and time "objectively" if

they are not the subjective forms of human intuition? And, second, why then do we need the classical conceptions of space and time in physics at all if they are merely our *subjective* forms of intuition, and do not correctly represent the objective structure? Let me take the second question first.

The answer, in principle, has already been given. We need the classical conception of space-time in order to design the bulk of physical experiments and to measure continuous quantities like distance, velocity etc., and in particular to separate *forces* from mere kinematical changes. However, the presupposition of the classical conception of space-time for the possibility of measurement is not an *absolute* one, as it is in Kant's transcendental epistemology. It is, rather, of a *contingent* and anthropomorphic nature, connected to human cognitive powers in the sense that we as *finite* beings can only recognize the world by making certain *ideal* assumptions regarding the structure of time and space. However,we have to bear in mind precisely that these *are* ideal assumptions. Thus, classical space-time is not presupposed as something *real*, but only as something *ideal*, in fact as a regulative principle (or idea) in Kant's sense. Once this distinction is made, any illusion as to the absoluteness of the presupposition vanishes. This in turn means that we must be prepared to *change* our ideal assumptions regarding space and time as and when this becomes *necessary* for the sake of physics, as was indeed the case at the turn of the century.

This brings me back to the first question regarding the *objective* structure of space and time. To this question, neither Hilbert nor anybody else has anything like a *final* answer: for any answer given will depend on the historical development, and the present state, of science. There can be no recognition of the objective structure of space and time without some ideal presuppositions, and these presuppositions may well turn out, in the light of developing science to be false, thus leading to further presuppositions, and so on.

Against this line of argument it could be objected that we are never *forced* to change an ideal assumption with respect to geometry, and that, in particular, Euclidean geometry can be maintained come what may. Poincaré's geometric conventionalism amounts to something like this, in particular that Euclidean geometry would have nothing to fear from future experience. It is, therefore, of interest to note that both Hilbert and Husserl (through his disciple Becker) vehemently opposed Poincaré's conventionalism. Hilbert's reasons can be briefly explained as follows. According to Helmholtz, our knowledge of geometry is essentially based on the notion of the *free mobility of rigid bodies* in space. But this, however, singles out only the spaces of 'constant curvature'. For this reason, Poincaré could quite correctly maintain that, provided space has constant curvature, the choice of an appropriate geometry is a matter of taste or convention, and that therefore the choice of Euclidean geometry is irrefutable by experiments with rigid bodies. However, according to Hil-

bert, the situation changes dramatically and fundamentally with the new
mechanics introduced by the special theory of relativity. Here the classical
conception of space and time has to be abandoned irrevocably because
Galileo's law of the additivity of velocities is *contradicted* by the body of
experiments available at the time. More precisely, the *impossibility* of
rigid bodies in the sense that there is no body which "has the same shape
and size when stationary and when in motion"[49] contradicts the Euclidean
conception of space as defended by Poincaré. This is so because if rigid
bodies existed *infinite* velocities would be possible, (something ruled out
by the special theory of relativity). Thus, for Hilbert, the *special* theory
of relativity *already* shows Poincaré's conventionalism to be wrong, and
this long before the advent of the general theory of relativity.

Let me return to Einstein's conception of the relationship between "geometry and experience", in particular his distinction of *two* kinds of geometry. As we have seen: Hilbert, too, differentiates between two kinds of
geometry. So far Hilbert and Einstein agree. However, they do not agree
on how and in what way the two geometries differ. In particular, they do
not agree on the epistemological way the distinction is to be justified.
Einstein suggests to divide geometry into a purely *logical-formal* geometry,
i.e., one not interpreted at all, in particular not intuitively, and a *practical*
geometry, i.e., one interpreted by coordinating real objects to geometrical
terms. Hilbert does *not* suggest such a division; in particular, he doesn't
assume that there is a pure logical-formal geometry as Einstein ascribes
to him, and indeed this would be very questionable from an epistemological point of view. Hilbert takes both kinds of geometry as interpreted,
but in a *different* way. The first, he thinks interpreted by recourse to
intuition anchored in the *Lebenswelt*; the second, becomes interpreted
through controlled experiments and precision measurements resulting in
a certain way from a refinement and an objectification of the intuition
rooted in the *Lebenswelt*. In the end, however, if we ignore the epistemologically different starting points, Hilbert and Einstein again agree that
geometry is a *natural* science based on real experiments and measurements. Thus, similarly to Einstein, Hilbert can assert:

Geometry is nothing but a branch of physics; in no way whatsoever do geometrical truths
differ essentially from physical truths nor are they of a different nature.[50]

NOTES

* I take this opportunity to thank my colleagues at ZiF for their very helpful comments and
criticisms during the academic year. For the preparation of this paper, I am especially
indepted to Michael Hallett, who not only improved my English but also saved me from a
number of errors.
[1] Einstein (1921, p. 3–4).

[2] Quine's: 'Two Dogmas of Empiricism' in *Philosophical Review* (1951) reprinted in *From a Logical Point of View*, Harper Torchbooks, New York.

[3] Einstein (1921, p. 4). The emphasis in the last sentence is mine; the expression 'Logical Formal' has been left almost the same as Einstein's original German term "das Logisch-Formale".

[4] This is Hilbert's own label for the "new" understanding of axioms, exemplified for the first time in his *Grundlagen der Geometrie* (1899). Whether the *Foundations of Geometry* really entail a "new" understanding of axioms [in oppositon to the traditional one] I will not discuss here. It suffices for my purpose to remark that the scientific community, in particular certain philosophers like Schlick, Carnap and Reichenbach, interpreted Hilbert's achievements in the *Grundlagen* very quickly in this particular way. Einstein is a further example of this trend, though, of course, a rather superb one.

[5] Einstein (1921, pp. 4–5).

[6] Indeed, Hilbert makes a clear distinction between *two* ways of recognising geometry, as we will see at the end of the paper. But this distinction is almost the contrary of Einstein's distinction between pure axiomatic and practical geometry.

[7] For the sake of clarity, I should add that the logical empiricists were not the only ones to interpret Hilbert in this formalist way: Frege did too. But there is an important difference between Frege and the empiricists: Whereas the latter agreed with this type of formalist reading, because it fitted so nicely with their epistemological views, Frege developed it in order to criticize Hilbert for his unclear notions of *axiom* and *definition*, and to reduce these to absurdity. Frege's polemic against Hilbert seems to me also unjustified and to rest on a dual misunderstanding. But I refrain from discussing this here.

[8] See Einstein (1921, pp. 5–6); the German word 'zuordnen' I have translated as 'coordination' although the latter does not capture precisely the sence of the action conveyed by the term "zuordnen".

[9] Hilbert (1891, p. 5ff).

[10] Hilbert (1891, pp. 6–7); Hilbert seems here to advocate a kind of 'logicicsm' with respect to arithmetic; this impression is, however, an illusion; it mainly results from the strong opposition of geometry and arithmetic; later he rejected the idea of logicism completely.

[11] Hilbert (1898a, p. 1); 'Geometrie ist die vollkommenste Naturwissenschaft'; see (1898).

[12] Hilbert (1898b, p. 2); 'Geometrie ist [dadurch] eine rein mathematische Wissenschaft [geworden]'; the words in brackets are later supplements.

[13] Hilbert (1899, p. 1) 'Die bezeichnete Aufgabe läuft auf die logische Analyse unserer räumlichen Anschauung hinaus'. See also (1898), (1898a) and (1902).

[14] See Hilbert (1898, p. 3ff).

[15] See Hilbert (1898, p. 5ff) and also (1891, p. 3ff).

[16] The ultimate reason is, of course, that Hilbert is immediately aware that the proposal of the logical empiricists, to split geometry into two parts, is no solution to the puzzle of geometry. On the contrary one must take geometry as a unity between mathematics and natural sciences and explain, why that is as it is, in order to solve the puzzle.

[17] See M. Friedman's essay in this volume. All three authors share the conviction that Euclidean geometry has a privileged position among the possible geometries; but they differ with respect to the reasons for its privilege. Carnap gives essentially an anthropomorphic explanation: the geometry of our intuition is indeed Euclidean, but it is valid only locally, in infinitesimally small regions, our close surroundings, so to speak. Weyl, on the other hand, gives a mathematical justification; he argues that the demand of continuous differentiability of the transformation functions for coordinates presupposes the validity of Euclidean metric in the 'infinitesimal small' (Weyl 1925, p. 12). Hilbert's justification of the privileged status of Euclidean geometry again differs from both.

[18] In particular I cannot offer a detailed account of the internal development of Hilbert's finitism, but have to take it in its mature form, as it was formulated for the first time in his 'Neubegründung der Mathemetik' (New Foundations of Mathematics) in 1922.

[19] Hilbert (1922, p. 162).

[20] Already in his 'Untersuchung über die Deutlichkeit der Grundsätze der natürlichen Theologie und Moral' (1764) Kant asserts: 'Mathematics considers in its solutions, proofs and consequences the general under the signs *in concreto*, philosophy [Weltweisheit] however, the general through the signs in abstracto' [AA 73]. But also later, in part II of the *Critique of Pure Reason* we find similar remarks: "Philosophy confines itself to universal concepts; mathematics can achieve nothing by concepts alone but hastens at once to intuition, in which it considers the concept in concreto" [B744/A716]. The last remark is particularly important because it shows that, according to Kant, there exists a very close connection between intuition and symbols in concreto. In intuition we consider a concept in concreto by constructing a concrete object which fits the concept, and in view of this object we prove the proposition in question.

[21] See the Inaugural Dissertation 'De mundi sensibilis atque intelligibilis', Section 12; compare this also with *Critique* B 49, where Kant explicitly states, "that the concept of change and with it the concept of motion [as change of position] is only possible in and by means of the representation of time": furthermore B 291/92 and *Prolegomena*, Section 10, as well as the letter to Schulz of 25 November, 1788.

[22] I use here a concept of explanation, which differs from the usual deductive-nomological conception of Hempel and Oppenheim, insofar as I judge it as essential that an explanation not only permits the logical deduction of singular propositions from law-like sentences, but also presents new and deeper reasons for the unity and order of a certain domain of facts and their objective relations to further domains. This "Tieferlegung der Fundamente" is, as Hilbert says, one of the main tasks of the axiomatic method.

[23] Hilbert (1922, p. 164); German original: "Bei der solcherart [anschaulich] betriebenen Zahlentheorie gibt es keine Axiome. und so sind auch keinerlei Widersprüche möglich. Wir haben eben konkrete Zeichen als Objekte. operieren mit diesen und machen über sie inhaltliche Aussagen.... Aber freilich läßt sich nicht die ganze Mathematik auf solche Art erfassen. Schon beim Übertritt zur höheren Arithmetik und Algebra, z. B. wenn wir Behauptungen über unendlich viele Zahlen oder Funktionen gewinnen wollen, versagt jenes inhaltliche Verfahren. Denn für unendlich viele Zahlen können wir nicht Zahlzeichen hinschreiben oder [geeignete] Abkürzungen einführen".

[24] The "tertium non datur" says in this case: Either all numbers have a certain property E, or there is at least one number, which has the property non-E. See Majer (1993, p. 60).

[25] Hilbert (1923, p. 181); [Italics are mine]: original in German: "Auf Grund der Axiome 1. bis 10. erhalten wir leicht die ganzen positiven Zahlen und die für diese geltenden Zahlgleichungen. Auch läßt sich aus diesen Anfängen mittels 'finiter' Logik durch rein anschauliche Überlegungen, wozu die Rekursion und die anschauliche Induktion für vorliegende endliche Gesamtheiten gehört, die elementare Zahlentheorie gewinnen, ohne daß dabei eine bedenkliche oder problematische Schlußweise zur Anwendung gelangt. Die beweisbaren Formeln, die auf diesem Standpunkt gewonnen werden haben sämtlich den Charakter des Finiten, d.h. die Gedanken. deren Abbilder sie sind, können auch *ohne irgendwelche Axiome* inhaltlich und unmittelbar mittels Betrachtung endlicher Gesamtheiten erhalten werden".

[26] *Critique* [A164/B204].

[27] "Die Arithmetik hat freylich keine Axiome, weil sie eigentlich kein Quantum (d.i. keinen Gegenstand der Anschauung als Größe) sondern blos die Quantität. d.i. einen Begriff von einem Ding überhaupt durch Größenbestimmung zum Objecte hat. Sie hat aber dagegen Postulate d.i. unmittelbar gewisse practische Urtheile . . . Das Urtheil $3 + 4 = 7$ scheint zwar ein blos theoretisches Urtheil zu sein . . . das + [bezeichnet aber] eine Art der Synthesis aus zwei gegebenen Zahlen eine dritte zu finden und eine Aufgabe, die keiner Auflösungsvorschrift noch eines Beweises bedarf. mithin ist das Urtheil ein Postulat.

[28] See Friedman (1992, p. xii), where he argues that Kant's philosophy (of geometry) is refuted precisely because it was so extraordinarily well adapted to the exact sciences of his days, in particular to Newton's physics. Compare. however. his contribution to this volume, where he considers the possibility of a modification of Kant's philosophy into a more general abstract form, which is compatible with the exact sciences of our century.

[29] The 'Festschrift' was Hilbert's contribution to a volume published in honor of Weber and Gauss which later became published separately under the title 'Grundlagen der Geometrie'. It was the result of a number of lectures on the foundations of geometry, that Hilbert had held since 1881. See Hilbert (1891, 1893, 1898, 1898a, 1899, 1902).

[30] This quotation however, must not be misinterpreted as a motto for the *Grundlagen* of the following sort: It was Hilbert's intention to *eliminate* intuition from geometry in favour of concepts. (This is suggested by H. Stein in his otherwise very interesting essay 'Logos, Logic, and Logistique'.) This would contradict, not only Hilbert's *fundamental* tendencies to base mathematics on a secure finite domain of intuitively identifiable objects, but also deprive the geometrical concepts of all their spatial meaning and, hence, reduce them to mere formal notions, an idea to which Hilbert has strongly opposed. See Hilbert 'On the Infinite' and my paper 'Different Forms of Finitism', in which I defend Hilbert's finite point of view against modern tendencies to dissolve Hilbert's finitism.

[31] Hilbert (1930): "Damit behält also der allgemeinste Grundgedanke der Kantschen Erkenntnistheorie seine Bedeutung: nämlich das philosophische Problem, jene anschauliche Einstellung a priori festzustellen und damit die Bedingungen der Möglichkeit jeder begrifflichen Erkenntis und zugleich jeder Erfahrung zu untersuchen", p. 961.

[32] Hilbert (1918, p. 3). In the original German: "Die genaue systematische Zusammenstellung der Eigenschaften des Raumes, die Untersuchung der zwischen ihnen bestehenden logischen Beziehungen und die Entwicklung der Konsequenzen aus ihnen bildet den Inhalt der Euklidischen Geometrie".

[33] It is important to note that Hilbert does not refer to *rigid bodies* and their motions as the proper object of geometry, as for example Helmholtz, Lie and Poincaré typically do. Instead, he speaks only of the "spatial positions and relations of things", and this quite from the beginning, long before the advent of the special theory of relativity. (See his lectures (1891, 1893 and 1898).) In this way, he avoids the difficulty to built geometry on a physical notion, namely that of rigid bodies, which is, strictly speaking, incompatible with special relativity, because, as Hilbert remarks: "there is no rigid body in the sense that it would have the same shape and size when motionless and when in motion" (1918, p. 82). See Majer (1995) where I explain Hilbert's critique of Poincaré's conventionalism more closely.

[34] Hilbert (1899, p. 2).

[35] Hilbert (1925, p. 190).

[36] Hilbert (1933).

[37] Hilbert (1933); the section closes with the sentence just quoted above.

[38] Corresponding sections for the practical validity of the classical conception of space and time can also be found in the early *Raum und Zeit* on pages 8 and 15:

> If this common view of space is valid, then naturally all the propositions of [Euclidean] geometry must be confirmable by experience. Thus, for example, the sum of the angles in a triangle must alvays be 180°. And this is indeed really the case.

And a bit later:

> The conception of time thus presented must – if it is applicable – prove itself in reality and, in fact, it does confirm itself brilliantly. The fact that the lives of different people can intermesh so intricately is based on its application.

[39] Husserl had left Göttingen in 1908 long before Hilbert gave his lectures on relativity and the latter were not circulated very much – at least to my knowledge.

[40] By "most of Hilbert's work on geometry" I do not only mean the 'Festschrift' and the other published papers but also the lecture series on geometry in 1898 and 1902 which Hilbert gave in Göttingen after Husserl had established his contacts with Hilbert.

[41] The geometry in question is that of Euclid.

[42] See (1891, p. 7): the original in German: "In der That entspringt denn auch die älteste Geometrie aus dem Anschauen der Dinge im Raum, wie sie das *tägliche Leben* bietet,und, wie alle Wissenschaft am Anfang, hat sie Probleme vom praktischen Bedürfnis gestellt".

[43] See (1898, p. 2); the original in German: "Diese [euklidische] Geometric ist gewisser-maßen die Geometrie des *täglichen Lebens*. Sie liegt aller Naturbetrachtung und aller Natur-wissenschaft zu Grunde".

[44] Friedman; this issue, pp. 255ff.

[45] Schmidt (in his contribution to this volume) points out that the assumption that space has a *differentiable*, structure is from a micro-physical perspective no less problematic than the assumption of rigid rods and homogenous clocks.

[46] It suffices here to concentrate on the special theory of relativity because the general theory of relativity was only a theoretical extension of certain aspects that had already been prepared in special relativity. Taken in this way, the special theory of relativity is the logical presupposition for the general theory.

[47] The complete answer as well as its justification is explained in Hilbert's two lectures on relativity, to which I have to refer the reader for any closer inspection.

[48] Hilbert (1921, p. 13); (italics by the author).

[49] Hilbert (1918, p. 82): see also Majer (1995) where Hilbert's criticism of Poincaré's conven-tionalism is developed in greater detail.

[50] Hilbert (1930, p. 962). There is a very similar remark in Einstein (1921, p. 6): "Geometry, thus supplemented, is obviously a natural science; we can definitely view it as the oldest branch of physics".

REFERENCES

Becker, O.: 1922, 'Beiträge zur phänomenologischen Begründung der Geometrie und ihrer physikalischen Anwendungen', *Jahrburgh für Philosophie* **VI**, 385–560.

Einstein, A.: 1921, *Geometrie und Erfahrung*, erweiterte Fassung des Festvortrages gehalten an der Preussischen Akademie der Wissenschaften zu Berlin, am 27. Januar 1921; Verlag Julius Springer, Berlin.

Friedman, M.: 1992, *Kant and the Exact Sciences*, Harvard University Press, Cambridge, Massachusetts.

Hilbert, D.: 1891, *Projektive Geometrie*, Vorlesung SS 1891.

Hilbert, D.: 1893, *Die Grundlagen der Geometrie*, Vorlesung; erstmals gehalten im SS 1884.

Hilbert, D.: 1898, *Grundlagen der Euklidischen Geometrie*; WS 1898/99.

Hilbert, D.: 1898a, *Elemente der Euklidischen Geometrie*, Vorlesungsmitschrift ausgearbeitet von E. Schaper.

Hilbert, D.: 1898b, *Mechanik*, Vorlesung WS 1898/99.

Hilbert, D.: 1899, *Grundlagen der Geometrie*, Festschrift zur Enthüllung des Gauss-Weber-Denkmales.

Hilbert, D.: 1902, *Grundlagen der Geometrie*, SS 1902, Vorlesungsnachschrift ausgearbeitet von A. Adler.

Hilbert, D.: *Raum und Zeit*, Vorlesung WS 1918/19 ausgearbeitet von P. Bernays, Mathema-tisches Institut der Universität Göttingen;.

Hilbert, D.: 1921, *Grundgedanken der Relativitätstheorie*, Vorlesung SS 1921; ausgearbeitet von P. Bernays, Mathematisches Institut der Universität Göttingen.

Hilbert, D.: 1922, *Neubegründung der Mathematik*.

Hilbert, D.: 1923, *Die logischen Grundlagen der Mathematik*.

Hilbert, D.: 1925, *Ueber das Unendliche*, in Gesammelte Werke, Bd. 2.

Hilbert, D.: 1930, 'Naturerkennen und Logik', *Naturwissenschaften* **18**.

Hilbert, D.: 1933, *Über das Unendliche*, Vortrag, Cod. MS 595 zitiert mit Erlaubnis der Niedersächsischen Staats- und Universitätsbibliothek Göttingen.

Husserl, E.: 1907, *Die Idee der Phänomenologie*, Fünf Vorlesungen by W. Biemel (ed.), Husserliana II, Martinus Nijhoff (1973).

Husserl, E.: 1907a, *Ding und Raum*, Vorlesungen 1907, by U. Claesges (ed.), Husserliana XVI, Martinus Nijhoff (1973).

Husserl, E.: 1913, *Ideen zu einer reinen Phänomenologie und Phänomenologischen Philosophie*, Husserliana III, (1950).[2]

Husserl, E.: 1936, *Die Krisis der Europäischen Wissenschaften und die Transzendentale Phänomenologie*, by W. Biemel (ed.), Husserliana VI, Martinus Nijhoff (1954).

Kant I.: 1764, *Untersuchung über die Deutlichkeit der Grundsätze der natürlichen Theologie und Moral*, Vorkritische Schriften, Akademie Ausgabe Bd. 1.

Kant I.: 1770, *De mundi sensibilis atque intelligibilis* . . . , Ak. Ausg. 3.

Kant I.: 1781, *Critique of Pure Reason*, translated by Kemp Smith.

Kant I.: 1783, *Prolegomena zu einer jeden künftigen Metyaphysik* . . . , Akademie Ausgabe Bd. 3.

Majer, U.: 1993, 'Hilberts Methode der idealen Elemente und Kants regulativer Gebrauch der Ideen', *Kant-Studien* **84**, 51–77.

Majer, U.: 1995, 'Hilbert's Criticism of Poincaré's Conventionalism', to be published in the Proceedings of the International Congress Henri Poincaré, Nancy, France.

Majer, U.: 1995a, 'Frege's Non-logical Basis of Arithmetic', in *Logic und Mathematik*, Jena.

Poincaré, H.: 1906, *Wissenschaft und Hypothese*, Teubner-Verlag, Leipzig.

Schlick, M.: 1925, *Allgemeine Erkenntnislehre*, STW 269, Frankfurt.

Stein, H.: 1988, 'Logos, Logic and Logistike', in W. Aspray and P. Kitcher (eds.), *History and Philosophy of Modern Mathematics*, University of Minnesota Press, Minnesota.

Weyl, H.: 1925, *Riemanns geometrische Ideen, ihre Auswirkung und ihre Verknüpfung mit der Gruppentheorie*, K. Chandrasekharan (ed.), Springer (1988).

Universität Hannover
Philosophisches Seminar
Postfach 6009
30060 Hannover
Germany